Targeted Muscle Reinnervation

Series in Medical Physics and Biomedical Engineering

Series Editors: John G Webster, E Russell Ritenour, Slavik Tabakov, Kwan-Hoong Ng, and Alisa Walz-Flannigan

Other recent books in the series:

Quantifying Morphology and Physiology of the Human Body Using MRI
L Tugan Muftuler (Ed)

Monte Carlo Calculations in Nuclear Medicine, Second Edition: Applications in Diagnostic Imaging
Michael Ljungberg, Sven-Erik Strand, and Michael A King (Eds)

Vibrational Spectroscopy for Tissue Analysis
M A Flower (Ed)

Webb's Physics of Medical Imaging, Second Edition
Ihtesham ur Rehman, Zanyar Movasaghi, and Shazza Rehman

Correction Techniques in Emission Tomography
Mohammad Dawood, Xiaoyi Jiang, and Klaus Schäfers (Eds)

Physiology, Biophysics, and Biomedical Engineering
Andrew Wood (Ed)

Proton Therapy Physics
Harald Paganetti (Ed)

Practical Biomedical Signal Analysis Using MATLAB®
K J Blinowska and J Żygierewicz (Ed)

Physics for Diagnostic Radiology, Third Edition
P P Dendy and B Heaton (Eds)

Nuclear Medicine Physics
J J Pedroso de Lima (Ed)

Handbook of Photonics for Biomedical Science
Valery V Tuchin (Ed)

Handbook of Anatomical Models for Radiation Dosimetry
Xie George Xu and Keith F Eckerman (Eds)

Fundamentals of MRI: An Interactive Learning Approach
Elizabeth Berry and Andrew J Bulpitt

Series in Medical Physics and Biomedical Engineering

Targeted Muscle Reinnervation
A Neural Interface for Artificial Limbs

Edited by

Todd A. Kuiken
Rehabilitation Institute of Chicago and Northwestern University, Illinois, USA

Aimee E. Schultz Feuser
Rehabilitation Institute of Chicago, Illinois, USA

Ann K. Barlow
Rehabilitation Institute of Chicago, Illinois, USA

CRC Press
Taylor & Francis Group
Boca Raton London New York

CRC Press is an imprint of the
Taylor & Francis Group, an **informa** business

A TAYLOR & FRANCIS BOOK

CRC Press
Taylor & Francis Group
6000 Broken Sound Parkway NW, Suite 300
Boca Raton, FL 33487-2742

First issued in paperback 2017

Library of Congress Cataloging-in-Publication Data

Targeted muscle reinnervation : a neural interface for artificial limbs / editors, Todd A. Kuiken, Aimee E. Schultz Feuser, Ann K. Barlow.
    pages ; cm. -- (Series in medical physics and biomedical engineering ; 28)
    Includes bibliographical references and index.
    ISBN 978-1-4398-6080-9 (hardcover : alk. paper)
    I. Kuiken, Todd A., editor of compilation. II. Feuser, Aimee E. Schultz, editor of compilation. III. Barlow, Ann K., editor of compilation. IV. Series: Series in medical physics and biomedical engineering ; 28.
    [DNLM: 1. Arm--innervation. 2. Nerve Regeneration--physiology. 3. Amputees--rehabilitation. 4. Artificial Limbs. WL 102]

    RD553
    617.5'7--dc23   2013018426

Visit the Taylor & Francis Web site at
http://www.taylorandfrancis.com

and the CRC Press Web site at
http://www.crcpress.com

To our patients, who continually inspire, amaze,

and challenge us to do better

Contents

About the Series

The Series in Medical Physics and Biomedical Engineering describes the applications of physical sciences, engineering and mathematics in medicine and clinical research.

The series seeks (but is not restricted to) publications in the following topics

- Artificial Organs
- Assistive Technology
- Bioinformatics
- Bioinstrumentation
- Biomaterials
- Biomechanics
- Biomedical Engineering
- Clinical Engineering
- Imaging
- Implants
- Medical Computing and Mathematics
- Medical/Surgical Devices
- Patient Monitoring
- Physiological Measurement
- Prosthetics
- Radiation Protection, Health Physics and Dosimetry
- Regulatory Issues
- Rehabilitation Engineering
- Sports Medicine
- Systems Physiology
- Telemedicine
- Tissue Engineering
- Treatment

The Series in Medical Physics and Biomedical Engineering is an international series that meets the need for up-to-date texts in this rapidly developing field. Books in the series range in level from introductory graduate textbooks and practical handbooks to more advanced expositions of current research.

The Series in Medical Physics and Biomedical Engineering is the official book series of the International Organization for Medical Physics.

The International Organization for Medical Physics

The International Organization for Medical Physics (IOMP), founded in 1963, is a scientific, educational, and professional organization of 76 national adhering organizations, more than 16,500 individual members, several Corporate Members, and four international Regional Organizations.

IOMP is administered by a Council, which includes delegates from each of the Adhering National Organizations. Regular meetings of Council are held electronically as well as every three years at the World Congress on Medical Physics and Biomedical Engineering. The President and other Officers form the Executive Committee and there are also committees covering the main areas of activity, including Education and Training, Scientific, Professional Relations, and Publications.

Objectives

- To contribute to the advancement of medical physics in all its aspects
- To organize international cooperation in medical physics, especially in developing countries
- To encourage and advise on the formation of national organizations of medical physics in those countries which lack such organizations

Activities

Official journals of the IOMP are *Physics in Medicine and Biology*, *Medical Physics*, and *Physiological Measurement*. The IOMP publishes a bulletin, *Medical Physics World*, twice a year that is distributed to all members.

A World Congress on Medical Physics and Biomedical Engineering is held every three years in cooperation with IFMBE through the International Union for Physics and Engineering Sciences in Medicine (IUPESM). A regionally based International Conference on Medical Physics is held between World Congresses. IOMP also sponsors international conferences, workshops, and courses. IOMP representatives contribute to various international committees and working groups.

The IOMP has several programs to assist medical physicists in developing countries. The joint IOMP Library Program supports 69 active libraries in 42 developing countries and the Used Equipment Program coordinates equipment donations. The Travel Assistance Program provides a limited number of grants to enable physicists to attend the World Congresses.

The IOMP website is being developed to include a scientific database of international standards in medical physics and a virtual education and resource center.

Information on the activities of the IOMP can be found on its website at www.iomp.org.

Preface

The concept of targeted muscle reinnervation (TMR) began almost three decades ago with a short line in a paper by Gerry Loeb and Andy Hoffer—one of those lines in the discussion where they were stretching for ideas. This line caught the eye of Todd Kuiken while he was searching for an idea for a PhD thesis. Years of animal research and mathematical modeling followed. Finally, feeling ready to try TMR in humans, Dr. Kuiken, a physiatrist and biomedical researcher, sought out Gregory Dumanian, MD, a plastics and hand surgeon. Over lunch, they hashed out many ideas and plans for implementing TMR in transhumeral amputees. Then a remarkable man named Jesse Sullivan came to the Rehabilitation Institute of Chicago with bilateral shoulder disarticulation amputations secondary to electrical burns. Jesse was not a transhumeral amputee, but he was the ideal patient/subject/research partner: he needed a revision surgery for painful shoulder scars, he was smart, he was articulate, he was compliant, and both Jesse and his wife Carolyn were such nice people to work with. Jesse consented to the procedure, and Dr. Dumanian developed a new plan and performed the first TMR surgery at the shoulder disarticulation level on Jesse. Six months later, when Jesse tried to close and open his hand, different parts of his chest muscle contracted, generating electrical signals that operated a prosthetic hand on the table—it worked! The team expanded: prosthetists Robert Lipschutz and Laura Miller developed novel methods of prosthetic fitting, with generous assistance from T. Walley Williams and Bill Sullivan of Liberating Technologies Inc. Kathy Stubblefield was the occupational therapist, with the patience of Job, who led the training. They all worked hard and skillfully together to deliver an amazing set of arms for Jesse to use with his new TMR control.

The multidisciplinary team approach has been essential in the evolution of TMR. The team has grown to include a host of different people: engineers of all kinds, neuroscientists, therapists, surgery residents, philanthropists, publicists, writers, and more. TMR has benefited from collaborations with many universities (in particular, with Kevin Englehart and his team at the University of New Brunswick) and with several industrial partners (in particular, Liberating Technologies Inc. and Otto Bock HealthCare GmbH). All of these collaborators were driven to make a clinical impact. At the heart of our science were our patients, who also became true research partners, giving great amounts of their time, their opinions, and such indomitable spirit.

We intend this book both as a template for clinical implementation of TMR and as a resource for research to further benefit the many thousands of individuals with amputations. We hope you will also build great multidisciplinary teams in order to implement TMR with your patients and to advance this new area of science.

A complementary website to this monograph, which contains videos of TMR surgical procedures, rehabilitation procedures, and new research developments, is available at www.RIC.org/targeted-muscle-reinnervation.

We gratefully acknowledge funding for this book and the complementary website from the National Institutes of Health, National Library of Medicine, award number G13LM011221.

Editors

Senior Editor

Todd A. Kuiken, MD, PhD, began studying nerve transfers to produce new electromyographic (EMG) signals while in graduate school, with the hope that these new signals could be used to improve the control of myoelectric prosthetic arms. After years of animal work followed by EMG simulation studies, the first human nerve transfer surgery for improved prosthesis control was performed in 2002. The technique, called targeted muscle reinnervation (TMR), was successful. Over the past 10 years, TMR has become an established clinical procedure, benefiting many patients across the United States and overseas. TMR has continued to evolve and improve, in particular with recent collaborative research on pattern recognition control. Dr. Kuiken has continued to lead efforts to understand and capitalize on the potential of TMR to provide improved prosthetic function. He leads an interdisciplinary team that includes physicians, prosthetists, therapists, neuroscientists, engineers, software developers, graduate students, and postdoctoral researchers at the Center for Bionic Medicine within the Rehabilitation Institute of Chicago. This combination of clinical and research expertise provides a unique environment in which to understand and develop TMR and to translate research data into clinical applications. Four integrated research groups within the Center for Bionic Medicine seek to study the functional and sensory benefits of TMR, to develop lighter, more functional prosthetic devices, and to design control systems to capitalize on the vast amount of neural information made available by TMR.

Dr. Kuiken earned a BS in biomedical engineering from Duke University, and a PhD in biomedical engineering and an MD from Northwestern University. He completed a residency in physical medicine and rehabilitation at the Rehabilitation Institute of Chicago and Northwestern University Medical School. Dr. Kuiken is the director of the Center for Bionic Medicine and the director of Amputee Services at the Rehabilitation Institute of Chicago. He is also a professor in the departments of Physical Medicine and Rehabilitation, Surgery, and Biomedical Engineering at Northwestern University. Dr. Kuiken is the recipient of many awards and honors for his work on TMR and is an internationally respected leader in the care of people with limb loss—both as a clinician and as a research scientist.

Associate Editors

Aimee E. Schultz Feuser, MS, has a BS in engineering from Swarthmore College, an MS in mechanical engineering from Northwestern University, and a certificate in editing from the University of Chicago. She began as a graduate student in the Center for Bionic Medicine in 2005 and continued with the team after finishing her MS, working first as a research engineer, and later, in addition, as a scientific writer. She has authored and coauthored several journal articles on targeted sensory reinnervation and prosthetics. In 2012 she became a freelance scientific writer and editor and continues to work with the Center for Bionic Medicine.

Ann K. Barlow, PhD, has a BSc in biochemistry and a PhD in molecular microbiology from the University of Southampton, UK, and an MS in written communication from National Louis University, Chicago. Dr. Barlow joined the Center for Bionic Medicine as a scientific writer in 2011. She supports all the research groups at the Center for Bionic Medicine with manuscript editing and grant development.

Contributors

Gregory A. Dumanian, MD, is chief and program director of the Division of Plastic Surgery at the Northwestern Feinberg School of Medicine, Chicago, and holds professorships in surgery, neurosurgery, and orthopedic surgery. Dr. Dumanian earned an AB in chemistry from Harvard University and an MD from the University of Chicago Pritzker School of Medicine. After a residency in general surgery at Massachusetts General Hospital, Dr. Dumanian completed plastic surgery training at the University of Pittsburgh and a fellowship in hand surgery at the Curtis Hand Center in Baltimore, Maryland. He is board certified in surgery, plastic surgery, and hand surgery. His specialties include state-of-the-art reconstructive breast surgery after cancer treatment, aesthetic surgery, abdominal wall reconstruction, peripheral nerve surgery, hand surgery, and reconstructive microsurgery. Dr. Dumanian co-developed the TMR procedure and performed the first TMR surgery in a human amputee in 2002.

Levi J. Hargrove, PhD, earned BSc, MSc, and PhD degrees in electrical engineering from the University of New Brunswick, Fredericton, New Brunswick, Canada. Since 2008, Dr. Hargrove has been director of the Neural Engineering for Prosthetics and Orthotics Lab at the Center for Bionic Medicine, Rehabilitation Institute of Chicago. He is also a research assistant professor in the Department of Physical Medicine and Rehabilitation and the Department of Biomedical Engineering at Northwestern University. His research interests include pattern recognition, biological signal processing, and myoelectric control of powered prostheses. Dr. Hargrove is a member of the Association of Professional Engineers and Geoscientists of New Brunswick.

Peter S. Kim, MD, earned an AB in human biology and an MD from Brown University's Program in Liberal Medical Education. After completing an integrated plastic and reconstructive surgery residency at Northwestern University, Dr. Kim completed a fellowship in hand and microvascular surgery at the Department of Orthopedics and Sports Medicine, University of Washington, Seattle. Dr. Kim is currently an instructor in surgery at Beth Israel Deaconess Medical Center at Harvard Medical School, where he specializes in hand, upper extremity, and reconstructive microsurgery. Dr. Kim has participated in numerous TMR procedures and has created a preclinical amputation model to investigate the effects of TMR on neuroma formation.

Jason H. Ko, MD, earned a BS in economics with minors in philosophy and chemistry and an MD from Duke University. Dr. Ko completed an integrated residency in plastic and reconstructive surgery at Northwestern Feinberg School of Medicine, Chicago, and a fellowship in hand and microvascular surgery at the University of Washington, Seattle. Dr. Ko is an assistant professor in the Division of Plastic Surgery, within the Department of Surgery, and in the Department of Orthopedics and Sports Medicine at the University of Washington, specializing in hand and upper extremity surgery, brachial plexus and peripheral nerve surgery, and reconstructive microsurgery. Dr. Ko has performed several preclinical animal studies examining the effects of TMR on neuroma formation and has performed numerous TMR procedures.

Robert D. Lipschutz, CP, earned a BS in mechanical engineering from Drexel University, Philadelphia, and a certificate in prosthetics and orthotics from the Post-Graduate Medical School, New York University. He completed his prosthetics training at the Shriners Hospital for Crippled Children in Springfield, Massachusetts and continued his clinical work at Newington Children's Hospital in Newington, Connecticut. He currently works as a research prosthetist for the Center for Bionic Medicine, is director of Prosthetics and Orthotics Education for the Prosthetics and Orthotics Clinical Center at the Rehabilitation Institute of Chicago, and is an assistant professor in the Department of Physical Medicine and Rehabilitation at Northwestern University. His current research interests include the fitting and evaluation of new advanced prosthetic devices.

Blair A. Lock, MS, earned BS and MS degrees in electrical engineering and a diploma in technology management and entrepreneurship from the University of New Brunswick, Fredericton, New Brunswick, Canada. Lock is a research engineer and the director of Research Operations at the Center for Bionic Medicine at the Rehabilitation Institute of Chicago. His research interests include pattern recognition for improved control of powered prostheses and user experience in rehabilitation technologies. Lock is a registered professional engineer with the Association of Professional Engineers and Geoscientists of New Brunswick, Canada.

Paul D. Marasco, PhD, earned a BA in biology from the University of Colorado, Colorado Springs, and a PhD in neuroscience from Vanderbilt University. Dr. Marasco is a principal investigator in the Advanced Platform Technology Center of Excellence and the director of Amputee Research in the Department of Physical Medicine and Rehabilitation at the Louis Stokes Cleveland Department of Veterans Affairs Medical Center. His research focuses on sensory integration with prosthetic devices within the context of systems-level mechanisms of brain organization, neural plasticity, and cognition.

Laura A. Miller, PhD, CP, earned a BS in biomedical engineering from Tulane University and MS and PhD degrees in biomedical engineering from Northwestern University, where she also obtained certification in prosthetics. She works as a research prosthetist for the Center for Bionic Medicine at the Rehabilitation Institute of Chicago and is an associate professor in the Department of Physical Medicine and Rehabilitation at Northwestern University. Her research interests include fitting and evaluation of new advanced prosthetic devices. Dr. Miller is a member of the International Society of Prosthetics and Orthotics and the American Academy of Orthotists and Prosthetists.

Douglas G. Smith, MD, earned his MD from the Pritzker School of Medicine, University of Chicago. Dr. Smith completed a residency in orthopedic surgery and rehabilitation at Loyola University Medical Center, Maywood, Illinois, and a fellowship in foot, ankle, and amputation surgery at Harborview Medical Center, University of Washington, Seattle. Dr. Smith is a professor in the Department of Orthopedics and Sports Medicine at the Harborview Medical Center and University of Washington, specializing in general orthopedic trauma and in surgery and rehabilitation for amputees, and has served as a consultant to the U.S. military amputee centers from 2002 to the present. Dr. Smith has performed more than 40 TMR procedures in acute, subacute, and chronic amputation settings, with a specific interest in the effects of TMR on pain.

Jason M. Souza, MD, earned a ScB in neuroscience from Brown University and an MD from Harvard Medical School. Dr. Souza is currently completing an integrated residency in plastic and reconstructive surgery at Northwestern Feinberg School of Medicine, Chicago, and is a lieutenant in the Medical Corps of the U.S. Navy. Dr. Souza has participated in numerous TMR procedures.

Kathy A. Stubblefield, OTR/L, earned a BS degree in occupational therapy from the University of Kansas. She works as a research therapist for the Center for Bionic Medicine at the Rehabilitation Institute of Chicago and has experience and expertise in the treatment of patients with stroke, spinal cord injury, and upper limb amputation. She has participated in research, training, and testing of TMR patients using conventional and experimental prosthetic components and control schemes.

1

Introduction

Todd A. Kuiken

Upper limb amputation causes a devastating disability. In the United States, an estimated 41,000 people live with arm amputation at or above the wrist (Ziegler-Graham et al. 2008); worldwide, this number is in the millions. The majority of unilateral and bilateral arm amputations are caused by trauma, such as industrial accidents, motor vehicle collisions, and—quite poignantly in this day and age—battlefield injuries. The next most common cause of arm amputation is dysvascular disease (Dillingham et al. 2002). Arm amputation can also be necessary in order to remove cancer, although improvements in cancer treatments and limb-sparing surgical techniques have made this less common. Finally, congenital limb deficiency, which occurs in 0.02% to 0.07% of live births (Ephraim et al. 2003), affects the upper limb in more than 50% of cases (Dillingham et al. 2002). Since trauma is the primary mechanism of upper limb amputation, the average age at which amputation occurs is quite young. Thus these individuals must live for decades with the consequences of losing a limb.

The functional impairment caused by arm amputation is clear: our hands are incredible tools. Losing a hand greatly limits what a person can do. Although people adapt and can live very full lives with only one intact arm, most tasks are harder, are more time-consuming, and frequently require a work-around. The loss of both hands is even more devastating and can make it extremely difficult to complete even the most basic tasks. The psychological impact of losing one or both arms is also enormous. Depression, anxiety disorders, and post-traumatic stress disorder are routinely seen, if not expected, in people who undergo upper limb amputation. Finally, losing an arm has a profound impact on social interactions, as we talk with our hands, greet one another with a handshake, and hold hands to show affection.

The loss of function caused by upper limb amputation is best treated using an appropriate prosthetic device. There are two general types of upper limb prostheses: body-powered and myoelectric. Body-powered devices were invented just after the Civil War and refined following World War I and World War II. Although advancements in materials, design, and manufacturing have been made, the body-powered devices worn today are very similar to the earliest patented design (Figure 1.1). Body-powered devices work

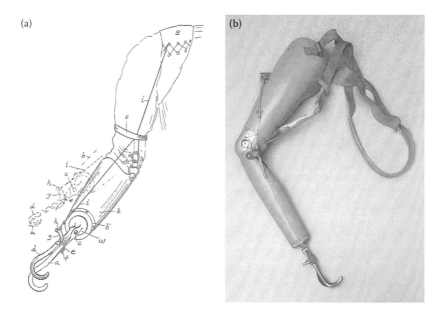

FIGURE 1.1
(a) Patent drawing for a body-powered prosthesis from 1912 (Dorrance 1912). (b) Modern body-powered prosthesis.

by harnessing shoulder motion: the patient protracts the shoulders to pull on a cable that transfers this movement to the prosthetic joints. A manual switch is generally used to change operation from one joint to another. In this manner, the patient can open or close a prosthetic hand or hook, then sequentially operate a wrist unit or an elbow. Body-powered devices are still popular today because they are robust and relatively simple to operate.

Myoelectric prostheses are motorized devices that are controlled using electromyographic (EMG) signals from residual muscles. These prostheses work fairly well for individuals with transradial amputations, where signals from the residual hand flexors and extensors in the forearm are used to open and close a prosthetic hand. This operation is intuitive because the remaining hand-control muscles are used to operate the prosthetic hand. However, hand-control muscles are lost with higher levels of amputation, as are muscles for controlling the wrist and, sometimes, the elbow. To use a myoelectric prosthesis, patients with transhumeral or shoulder disarticulation amputations must therefore use the very unnatural process of activating their upper arm or chest muscles to control the prosthetic wrist or hand.

Unfortunately, current upper limb prostheses for high-level amputees are cumbersome and awkward to control and do not adequately restore lost function. This is perhaps best illustrated by the fact that many unilateral upper limb amputees choose not to use a prosthesis at all: the benefits of these devices are outweighed by the poor control and discomfort of wearing a prosthesis (Biddiss and Chau 2007). However, most bilateral arm amputees must use a prosthesis in order to regain any lost function. Interestingly, bilateral amputees more commonly use body-powered devices because (1) they can be operated more quickly, (2) they enable both wrist rotation and wrist flexion/extension, and (3) they are more robust than myoelectric devices.

A variety of robotic arms are available that include up to three motorized joints: powered elbows, powered wrist rotators, and powered hands that open and close (Figure 1.2). Prosthetic hands with individually actuated fingers have recently become available. Highly articulated prosthetic limbs with up to 20 degrees of freedom have also been developed in research settings but are not yet commercially available. The challenge in using all of these devices—especially for high-level amputees—is to intuitively control the multiple available movements.

To control a multifunctional prosthesis, a neural interface is needed that can record motor commands intended for the missing limb, decipher the user's intended movements, and command the prosthetic arm. Such an interface could allow intuitive control of more complex devices, as prosthesis users would simply use their natural motor control commands to operate relevant prosthetic joints. In addition, providing closed-loop control by

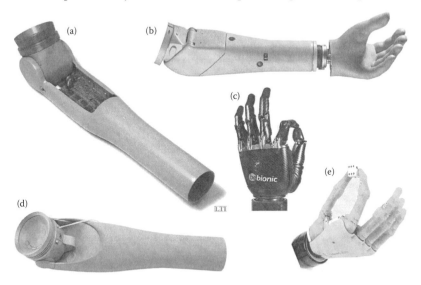

FIGURE 1.2
Currently available prosthetic systems: (a) Boston Digital Arm System from Liberating Technologies Inc. (Image provided compliments of Liberating Technologies Inc.) (b) Utah Arm from Motion Control Inc. (Image provided compliments of Motion Control Inc.) (c) Bebionic3 Hand from SteeperUSA. (Image provided compliments of SteeperUSA.) (d) DynamicArm® (Otto Bock Healthcare GmbH, Duderstadt, Germany) from Otto Bock. (Image provided compliments of Otto Bock.) (e) i-limb™ ultra from Touch Bionics® (Touch Bionics Inc., Mansfield, MA). (Image provided compliments of Touch Bionics.)

returning relevant sensory signals to the user would correlate prosthesis activity with limb sensation, resulting in an important feedback system, and improved control.

Two of the most obvious places to create an interface with the human nervous system are the brain (*brain-machine interfaces*) and the peripheral nerves (*peripheral nerve interfaces*). Brain-machine interfacing has become a very large and exciting field in recent years. Using arrays of electrodes that penetrate the cortex, many laboratories have demonstrated the ability to record control signals and operate devices (Taylor et al. 2002; Hochberg et al. 2006; Velliste et al. 2008). Most of the research to date has been in animal models, including a primate model in which monkeys were able to control three-degree-of-freedom robotic arms for feeding (Velliste et al. 2008). In the few studies that have been performed in humans, cortical arrays were implanted into severely disabled patients (with high spinal cord injury, brainstem stroke, or amyotrophic lateral sclerosis [ALS]). The patients could then use brain waves to control the cursor on a computer screen, do object-tracking tasks, use environmental control systems, or control robotic arms (Kubler et al. 2005; Hochberg et al. 2006; Kim et al. 2008; Hochberg et al. 2012). High-density electroencephalography (EEG) has also been used to record both intended speech and motor commands to operate simple spelling devices and perform basic movements, respectively (Birbaumer et al. 1999; Arbel et al. 2007; Iturrate et al. 2009).

The development of systems that provide sensation through brain-machine interfaces is still in its infancy. Some research has suggested that sensory data can be supplied through the somatosensory cortex (Romo et al. 1998; Romo et al. 2000). It is also well known that areas of the brain can be stimulated during craniotomy procedures and that patients will feel unique sensations or hear sounds (Vignal et al. 2007; Mulak et al. 2008).

Developing a neural-machine interface for amputees by accessing peripheral nerves is perhaps more logical. Peripheral nerves carry the centrally processed and integrated motor commands for complex intended movements to individual effector muscles. In addition, stimulation of afferent nerves provides defined sensations that can be localized to specific skin regions (Dhillon et al. 2004). Although research on direct peripheral nerve interfacing has been ongoing for more than 50 years, it has not yet been clinically implemented.

There are significant challenges facing the clinical implementation of both brain-machine interfaces and peripheral nerve interfaces. The signals that must be recorded are very small, on the order of microvolts, and peripheral-nerve signals are inundated by surrounding EMG signal noise with a similar bandwidth. Furthermore, it is very difficult to keep microelectrodes stable and in the correct position for long periods. Over time, the nervous tissue will scar and after a few months the implants will generally fail. Finally, these systems are fragile and require surgical replacement.

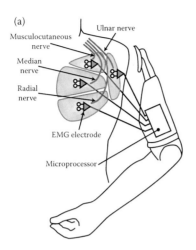 

FIGURE 1.3
(a) Schematic of targeted reinnervation for a shoulder disarticulation patient, showing transfers of four brachial plexus nerves for myoelectric prosthesis control. (Modified from Stubblefield et al., Occupational therapy protocol for amputees with targeted muscle reinnervation, *J Rehabil Res Dev* 46 (4):481–488, 2009. With permission.) (b) Shoulder disarticulation patient performing a functional test after TMR (using an experimental prosthesis. Miller et al. 2008.)

Targeted muscle reinnervation (TMR) is a new approach to recording signals from peripheral nerves and providing sensory feedback to the prosthesis user. During TMR surgery, residual arm nerves are transferred to spare *target* muscles in the residual limb of transhumeral amputees or the thorax of shoulder disarticulation amputees. Target muscles are denervated and the residual nerves are coapted to the remaining motor points. Once reinnervation of the target muscles is complete, the muscles will contract in response to activation of the transferred nerves, producing EMG signals that can be recorded from the overlying skin and used to control a prosthesis (Figure 1.3). Essentially, the target muscle acts as a *biological amplifier* of the peripheral nerve signals. The EMG signals resulting from contractions of the target muscles can be used to control analogous movements of the prosthesis, resulting in intuitive control. For example, after TMR, when a shoulder disarticulation amputee attempts to bend his or her missing elbow, a command signal travels down the musculocutaneous nerve to the new target muscle (e.g., a segment of pectoralis major). This muscle segment contracts and generates an EMG signal that can be used to flex the elbow of the prosthetic arm. Four major brachial plexus nerves innervate the arm and hand; thus TMR can produce at least four new control signals in shoulder disarticulation patients and up to five unique signals in transhumeral amputees (including those from remaining intact nerve-muscle connections).

During TMR surgery, sensory afferents in the transferred nerves can also be directed to reinnervate nearby skin. When this reinnervated skin is touched, patients feel as though their missing hand or arm is being touched; we call this phenomenon *transfer sensation*. Patients can feel light touch, graded pressure, different frequencies of vibration, and hot and cold stimuli—all as if these stimuli were applied to their missing hand (Kuiken et al. 2007; Sensinger et al. 2009; Schultz et al. 2009). These true physiologic sensations correlate with both the anatomy of the hand and the type of stimulation; thus there is no sensory substitution involved. Transfer sensation is unrelated to phantom limb sensation—the perception of the missing limb in space—which is poorly understood but is likely centrally mediated (Flor et al. 2000). In fact, patients often need to be trained to ignore phantom limb sensation as they learn to intuitively control their prosthesis.

TMR is a very pragmatic approach to improving the control of artificial limbs and has many advantages over other neural-machine interfaces. The surgical procedure for nerve transfer—detailed in Chapter 3—is very successful and provides a permanent neural interface. No implanted hardware or batteries are required, as the muscles intrinsically provide the energy to generate the control signals. Finally, commercially available prosthetic arm systems, with minor changes to software and incorporation of additional electrodes, can be used.

This book is intended to be a practical reference for implementing TMR—a guide for the entire multidisciplinary team. Chapter 2 describes the basic scientific concepts and key principles underlying TMR. Surgical approaches for individuals with transhumeral and shoulder disarticulation amputations are then described in detail in Chapter 3 (a supplemental training video on TMR surgery for transhumeral amputation is available on the Center for Bionic Medicine [CBM] website). A possible role for TMR in the prevention and treatment of end-neuromas is discussed in Chapter 4. Chapters 5, 6, and 7 detail the basic principles of rehabilitation, prosthetic fitting, and occupational therapy for the TMR patient. Chapter 8 provides the details of transfer sensation and describes the potential for providing sensory feedback to amputees. The surgical and functional outcomes for the first several TMR patients are presented in Chapter 9. The final chapter presents some exciting research that we believe will move TMR even closer to the goal of improving the function and quality of life for people with limb loss.

We hope you find this to be a useful guide for implementing TMR in patients with high-level upper limb amputations. We also hope this guide serves as a foundation to enable improvements in TMR techniques and advances in prosthetic technology. There is much more research to be done and much, much more room to improve the control and function of artificial limbs.

References

Arbel, Y., R. Alqasemi, R. Dubey, and E. Donchin. 2007. Adapting the P300-brain computer interface (BCI) for the control of a wheelchair-mounted robotic arm system. *Psychophysiology* 44:S82–S83.

Biddiss, E., and T. Chau. 2007. Upper-limb prosthetics: Critical factors in device abandonment. *Am J Phys Med Rehab*, 86 (12):977–987.

Birbaumer, N., N. Ghanayim, T. Hinterberger, et al. 1999. A spelling device for the paralysed. *Nature* 398 (6725):297–298.

Dhillon, G. S., S. M. Lawrence, D. T. Hutchinson, and K. W. Horch. 2004. Residual function in peripheral nerve stumps of amputees: implications for neural control of artificial limbs. *J Hand Surg [Am]* 29A (4):605–615.

Dillingham, T. R., L. E. Pezzin, and E. J. MacKenzie. 2002. Limb amputation and limb deficiency: epidemiology and recent trends in the United States. *South Med J* 95 (8):875–883.

Dorrance, D. W. 1912. Artificial Hand U.S. Patent 1,042,413, filed February 17, 1912, and issued October 29, 1912.

Ephraim, P. L., T. R. Dillingham, M. Sector, L. E. Pezzin, and E. J. MacKenzie. 2003. Epidemiology of limb loss and congenital limb deficiency: a review of the literature. *Arch Phys Med Rehabil* 84 (5):747–761.

Flor, H., W. Muhlnickel, A. Karl, et al. 2000. A neural substrate for nonpainful phantom limb phenomena. *Neuroreport* 11 (7):1407–1411.

Hochberg, L. R., D. Bacher, B. Jarosiewicz, et al. 2012. Reach and grasp by people with tetraplegia using a neurally controlled robotic arm. *Nature* 485:372–375.

Hochberg, L. R., M. D. Serruya, G. M. Friehs, et al. 2006. Neuronal ensemble control of prosthetic devices by a human with tetraplegia. *Nature* 442 (7099):164–171.

Iturrate, I., J. M. Antelis, A. Kubler, and J. Minguez. 2009. A noninvasive brain-actuated wheelchair based on a P300 neurophysiological protocol and automated navigation. *IEEE Trans Robotics* 25 (3):614–627.

Kim, S. P., J. D. Simeral, L. R. Hochberg, J. P. Donoghue, and M. J. Black. 2008. Neural control of computer cursor velocity by decoding motor cortical spiking activity in humans with tetraplegia. *J Neural Eng* 5 (4):455–476.

Kubler, A., F. Nijboer, J. Mellinger, et al. 2005. Patients with ALS can use sensorimotor rhythms to operate a brain-computer interface. *Neurology* 64 (10):1775–1777.

Kuiken, T. A., P. D. Marasco, B. A. Lock, R. N. Harden, and J. P. A. Dewald. 2007. Redirection of cutaneous sensation from the hand to the chest skin of human amputees with targeted reinnervation. *Proc Natl Acad Sci U S A* 104 (50):20061–20066.

Miller, L. A., R. D. Lipschutz, K. A. Stubblefield, et al. 2008. Control of a six degree of freedom prosthetic arm after targeted muscle reinnervation surgery. *Arch Phys Med Rehabil* 89 (11):2057–2065.

Mulak, A., P. Kahane, D. Hoffmann, L. Minotti, and B. Bonaz. 2008. Brain mapping of digestive sensations elicited by cortical electrical stimulations. *Neurogastroenterol Motil* 20 (6):588–596.

Romo, R., A. Hernandez, A. Zainos, C. D. Brody, and L. Lemus. 2000. Sensing without touching: psychophysical performance based on cortical microstimulation. *Neuron* 26 (1):273–278.

Romo, R., A. Hernandez, A. Zainos, and E. Salinas. 1998. Somatosensory discrimination based on cortical microstimulation. *Nature* 392 (6674):387–390.

Schultz, A. E., P. D. Marasco, and T. A. Kuiken. 2009. Vibrotactile detection thresholds for chest skin of amputees following targeted reinnervation surgery. *Brain Res* 1251:121–129.

Sensinger, J. W., A. E. Schultz, and T. A. Kuiken. 2009. Examination of force discrimination in human upper limb amputees with reinnervated limb sensation following peripheral nerve transfer. *IEEE Trans Neural Sys Rehabil Eng* 17 (5):438–444.

Stubblefield, K. A., L. A. Miller, R. D. Lipschutz, and T. A. Kuiken. 2009. Occupational therapy protocol for amputees with targeted muscle reinnervation. *J Rehabil Res Dev* 46 (4):481–488.

Taylor, D. M., S. I. Tillery, and A. B. Schwartz. 2002. Direct cortical control of 3D neuroprosthetic devices. *Science* 296 (5574):1829–1832.

Velliste, M., S. Perel, M. C. Spalding, A. S. Whitford, and A. B. Schwartz. 2008. Cortical control of a prosthetic arm for self-feeding. *Nature* 453 (7198):1098–1101.

Vignal, J. P., L. Maillard, A. McGonigal, and P. Chauvel. 2007. The dreamy state: hallucinations of autobiographic memory evoked by temporal lobe stimulations and seizures. *Brain* 130:88–99.

Ziegler-Graham, K., E. J. MacKenzie, P. L. Ephraim, T. G. Travison, and R. Brookmeyer. 2008. Estimating the prevalence of limb loss in the United States: 2005 to 2050. *Arch Phys Med Rehabil* 89 (3):422–429.

2

The Scientific Basis of Targeted Muscle Reinnervation

Todd A. Kuiken

CONTENTS

2.0 Introduction

After an amputation, severed residual nerves continue to transmit motor control information intended for the missing limb (Gordon et al. 1980). This neural activity may persist for decades, if not indefinitely (Davis et al. 1978; Dhillon et al. 2004; Jia et al. 2007). In addition, severed nerves can successfully reinnervate nonnative muscles (Elsberg 1917), and these reinnervated muscles subsequently respond appropriately to nonnative nerve commands (Gordon et al. 1980; O'Donovan et al. 1985; Gordon et al. 1986). These key observations provided the basis for the development of targeted muscle reinnervation (TMR) (Kuiken 2003; Kuiken et al. 2004). TMR critically depends on two key elements: (1) robust reinnervation of target muscle and (2) generation of strong, independent electromyographic (EMG) signals. In this chapter, we review the scientific and physiologic principles that underlie the successful implementation of TMR in human subjects and provide the foundation for further development of this technique.

2.1 Principles of Targeted Muscle Reinnervation

2.1.1 Hyper-Reinnervation of Muscle

When a motor neuron is cut, the distal part of the nerve undergoes Wallerian degeneration (Waller 1850). The proximal fiber attempts to regenerate by sending out neural processes that have the potential to form new functional connections with muscle fibers (i.e., to *reinnervate* the muscle). However, muscle often does not regain previous mass or functional capacity after reinnervation (Frey et al. 1982). This may be due to a failure of the regenerating fiber to reinnervate all available muscle fibers and/or attempts to reinnervate incompatible targets (e.g., attempted reinnervation of a muscle fiber by a sensory nerve) (Madison et al. 2007). A muscle reinnervated by its original nerve will, over time, demonstrate near-normal motor unit properties, axonal conduction speeds, muscle contractile speeds, and fatigue resistance (Foehring et al. 1986). However, self-reinnervated muscles regain only 70% to 80% of function (Frey et al. 1982). They also demonstrate a decline in mean maximum tetanic tension (Foehring et al. 1986) and an increase in adipose and connective tissue (Frey et al. 1982). Despite advances in microsurgical techniques, obtaining consistently good functional results after surgical repair of peripheral nerve damage, especially after nerve transection, is still a challenge (Scholz et al. 2009; Siemionow and Brzezicki 2009).

Reinnervation of a denervated muscle with an excess of donor nerve fibers (i.e., *hyper-reinnervation*) was shown to result in improved muscle recovery, in terms of muscle mass and strength, in a rat reinnervation model (Kuiken et al. 1995). In this model, rat medial gastrocnemius muscle was denervated and reinnervated with either the original medial gastrocnemius (mg) nerve (self-reinnervation) or one of four different combinations of additional nerves that provided up to 12 times the number of nerve fibers found in the original mg nerve. The degree of hyper-reinnervation corresponded to the degree of muscle recovery: muscles hyper-reinnervated by multiple nerves recovered muscle mass and muscle strength more fully than self-reinnervated muscles or muscles hyper-reinnervated with fewer nerves (Figure 2.1). When reinnervated with approximately 12 times the native number of nerve fibers, relative muscle mass recovered to 94.4% (SD 8.2%) of normal levels. This was significantly indistinguishable from contralateral, non-denervated controls and significantly greater ($p < 0.005$) than the 67.6% (SD 24.7%) recovery achieved by self-reinnervation. In addition, the increased variability in the extent of muscle recovery after self-reinnervation compared to hyper-reinnervation

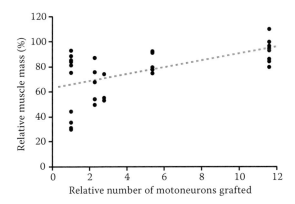

FIGURE 2.1
Relative recovery (by mass) of denervated rat medial gastrocnemius muscle after reinner-vation with increasing numbers of donor nerve fibers. The relative number of motoneurons grafted equals the estimated number of motoneurons contained in various grafts divided by the number of motoneurons in the mg nerve. Dots indicate separate animals. The correlation coefficient of the regression line is 0.60. (Adapted from Kuiken et al., The hyper-reinnervation of rat skeletal muscle, *Brain Res* 676:113–123. © [1995]. With permission from Elsevier.)

indicates that muscle recovery is also more consistent after hyper-reinner-vation. These results suggest that hyper-reinnervation greatly increases the probability of any given motor end plate being reinnervated, thus ensuring robust reinnervation of the muscle. In addition, hyper-reinnervation resulted in the formation of many more functional motor units: reinnervation with approximately 12 times the normal number of nerves fibers resulted in for-mation of up to three times as many functional motor units as in normal control muscle.

In TMR, large (8–10 mm) brachial plexus nerves containing thousands of motoneurons are transferred onto motor points of small (~1 mm) dis-tal motor nerves that innervate relatively small regions of muscle, greatly increasing the odds of any individual muscle fiber being reinnervated. Thus, excellent recovery of the target muscle is expected after TMR. The ability to form additional functional motor units in hyper-reinnervated muscle has important implications for TMR, as it suggests that target muscles have the capacity to be reinnervated by many more motoneurons. The large brachial plexus nerves innervate many small muscles of the hand, wrist, and arm and carry a vast number of separate control signals. Reinnervated muscles would thus provide a rich source of EMG control information that could be used to restore additional functions of the miss-ing arm and hand. High-density EMG studies of reinnervated muscles in fact demonstrate that reinnervated muscle contains control information for individual finger and thumb movements (Zhou et al. 2007). (See Chapter 10 for future methods of utilizing this information.)

2.1.2 Long-Term Viability of Severed Nerves

In a rat model of prolonged axotomy prior to reinnervation, chronically severed neurons could regenerate into and reinnervate a freshly denervated,
foreign muscle (Fu and Gordon 1995). Although the number of axons that
were capable of regenerating decreased over time, axon branching and an
increase in motor unit size resulted in near-normal muscle size and strength.
Motor activity declines in severed nerves; however, substantial recovery
occurs after reinnervation of muscle targets (Gordon et al. 1980).

   In human amputees, TMR has been performed many years after the original amputation and severing of brachial plexus nerves. The time limit for successful reinnervation is unknown. However, given the vast excess of nerve
fibers transferred to healthy, normal muscle in TMR, even if some motor
neurons are lost with time or are prevented from making functional connections by scar tissue, reinnervation of the newly denervated target muscle is
likely to be robust. TMR has been performed up to 6 years post amputation,
with good results.

2.1.3 Denervation of Target Muscle

For successful conventional myoelectric control after TMR, each reinnervated muscle or muscle segment must generate an independent EMG signal (i.e., a signal with no interference from other muscle EMG signals).
The most important factor in ensuring independent signals is complete
surgical denervation of the target muscle during TMR. Many muscles
are innervated by multiple motor nerve branches. The best example is
the pectoralis major muscle, which is innervated by two separate nerves
that branch so that there are usually three distinct sites where the nerve
penetrates the muscle (i.e., motor points), although there can be additional motor points in some individuals. The nerves to the lateral triceps
branch in a more variable pattern, often forming several motor points.
The heads of the biceps often have a single motor point; however, there
are other variations. If all native nerve branches to a target muscle are
not cut during surgery, then that target muscle will have a combination
of natively innervated muscle fibers and reinnervated muscle fibers. This
can greatly interfere with using the target muscle for myoelectric control
because the native nerve(s) will generate unwanted EMG signals. Failure
to fully denervate a target muscle may also prevent robust reinnervation
because fewer motor end plates are available to the transferred nerve
and/or because remaining native nerves will compete for reinnervation
of available sites (Hoffman 1950; Thompson and Jansen 1977). Thus it is
important that the surgeon carefully find and cut all of the nerve branches
that innervate the target muscle.

2.1.4 Independent Reinnervation of Adjacent Muscles/Muscle Segments

Each brachial plexus nerve controls a discrete set of hand and arm movements and, as such, can provide independent control signals for different prosthetic functions. Thus a key requirement in TMR is that each nerve reinnervates an independent, separate muscle (or muscle segment). In shoulder disarticulation amputees, large brachial plexus nerves with an excess of actively reinnervating axons are transferred onto adjacent segments of a single muscle, for example, pectoralis (Kuiken et al. 2004). These regenerating axons actively compete to reinnervate any available motor end plate. Because large numbers of nerve fibers are transferred to relatively small muscle segments, it is possible that nerves could grow into and reinnervate adjacent muscle segments, although transfer of a nerve to a muscle motor point likely provides a competitive advantage to that nerve for reinnervation of that muscle (Elsberg 1917). Such *cross-reinnervation* of adjacent muscles or muscle segments would result in mixed EMG signals, preventing separation of discrete control signals. Cross-reinnervation can be prevented by surgically separating muscle segments to generate a physical barrier of scar tissue (Kuiken et al. 2004). Alternatively, an adipofascial flap (O'Shaughnessy et al. 2008) can be inserted to physically separate muscles/muscle segments. (See Chapter 3 for description of this surgical procedure.)

2.2 Optimization of the EMG Signal

Optimal conventional control of a myoelectric prosthesis requires EMG signals that are strong (i.e., of sufficient amplitude to be detected and separated from background noise) and free of cross talk (i.e., not contaminated by EMG signals from adjacent muscles). Because TMR creates additional EMG control sites that are frequently located within a limited surface area, the ability to distinguish each signal becomes a more important issue.

To understand the parameters necessary for increasing signal strength and reducing cross talk from reinnervated muscles, theoretical finite element (FE) models were used (e.g., as shown in Figure 2.2) that included the structural components of a residual limb: bone, muscle, fat, and skin. Studies were performed to identify parameters that affect EMG signal propagation, signal size, and signal independence.

2.2.1 Signal Strength

The key issues for the strength of the EMG signal are the size of the muscle and the depth of the overlying subcutaneous fat (Lowery et al. 2002; Lowery et al. 2003; Kuiken et al. 2003). Muscle fibers near the surface contribute most

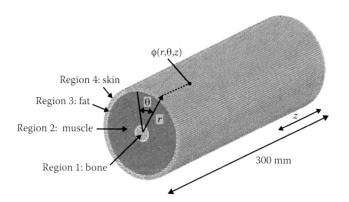

FIGURE 2.2
Cross-sectional view of finite element (FE) model of a transhumeral residual limb. This model includes skin, subcutaneous fat, muscle and a central bone. (Adapted from Lowery et al., A multiple-layer finite-element model of the surface EMG signal, *IEEE Trans Biomed Eng* 49(5):446–454. © [2002] IEEE. With permission.)

of the power to an EMG signal. With subcutaneous fat layers of 0 to 3 mm, 95% of the signal power is contributed by the first 6 to 15 mm of muscle (Lowery et al. 2003). Thus target muscle should ideally be at least 1 cm thick. The width of the muscle ideally should be bigger than the standard electrode spacing, which is generally 2 to 3 cm. Thus the target muscle surface should be larger than this, preferably at least 3 to 5 cm wide (Figure 2.3).

The subcutaneous fat lying over muscle acts as a spatial filter, greatly attenuating the surface EMG signal and reducing the focus of the recording area. Reducing the depth of the fat layer from 18 mm (the thickness typically found over the chest or outer upper arm) to 3 mm dramatically increases the amplitude of the EMG signal, by a factor of about 5 (Kuiken et al. 2003). These results agree with previous experimental observations that showed a

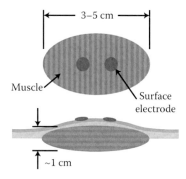

FIGURE 2.3
(See color insert.) Minimum size of target muscle required for detection of adequate EMG signal strength after TMR.

reduction of EMG signal with increased subcutaneous fat (Solomonow et al. 1994) and increased skinfold thickness (De la Barrera and Milner 1994).

Bone can also affect EMG signal magnitude. FE studies indicated that bone located close to the skin surface can cause significant disruption in the propagation of EMG signals (Lowery et al. 2002). Bone reduces conduction currents and results in a significantly increased signal on one side of the bone and a reduced signal on the other side. This is likely of little importance in transhumeral amputees in whom the bone is essentially centrally located. However, in shoulder disarticulation patients, nerve transfers to the clavicular head of the pectoralis muscle are frequently performed during TMR. This muscle is adjacent to the clavicle bone, which is close to the skin surface and thus generates very strong EMG signals.

2.2.2 Signal Separation

As the thickness of the subcutaneous fat is reduced, the EMG signal decays more rapidly as the electrode is moved farther from the source around the limb (i.e., with increasing angular displacement of the electrode with respect to the source) (Figure 2.4a). Similarly, the pickup volume (the volume of muscle that contributes 95% of the surface EMG signal) decreases as subcutaneous fat thickness decreases (Figure 2.4b). The rate of decay of the

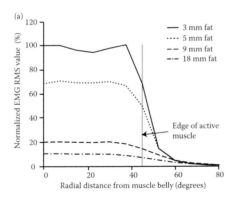 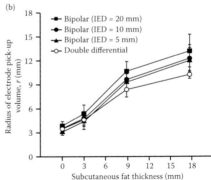

FIGURE 2.4
(a) Rate of decay of surface EMG signal as a function of radial distance from active muscle belly calculated using FE analysis. The amplitude of the surface EMG signal is normalized with respect to the signal calculated for an electrode located directly above the active muscle, with no subcutaneous fat. At zero degrees, the electrode is located directly above the active fiber. (Adapted and reprinted from Kuiken et al., The effect of subcutaneous fat on myoelectric signal amplitude and cross-talk. *Prosthet Orthot Int* 27(1):48–54. © [2003] Rehabilitation Institute of Chicago, Chicago, IL. With permission from SAGE.) (b) Influence of subcutaneous fat on pickup volume of surface electrode (using bipolar electrodes with various inter-electrode distances [IED]). (Reprinted from Lowery et al., Independence of myoelectric control signals examined using a surface EMG model, *IEEE Trans Biomed Eng* 50 (6):789–793. © [2003] IEEE. With permission.)

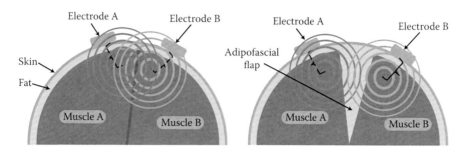

FIGURE 2.5
(See color insert.) Use of an adipofascial flap to physically separate adjacent muscles or muscle segments. *Left*: Cross talk is caused by interference of EMG signal from adjacent muscles: electrode A (on muscle A) picks up EMG signals from muscle B, and electrode B picks up EMG signals from muscle A. *Right*: Physical separation of muscles A and B by an adipofascial flap reduces the cross talk detected at both electrodes A and B. EMG signals (*concentric blue circles*) are shown propagating from a point source, for clarity.

surface potential with distance, and the size of the pickup volume, affects the amount of cross talk. A higher rate of signal decay (and a smaller pickup volume) results in less cross talk and allows signals that are relatively close together to be distinguished. The FE data suggest that when there is little or no fat between the electrode and muscle, EMG signals from superficial muscles, lying 5 to 10 mm below the surface of the muscle tissues, can be independently detected (Lowery et al. 2003).

Subcutaneous fat is therefore removed over the target muscles during TMR surgery to maximize signal strength and minimize cross talk. If subcutaneous fat is also removed from nontarget muscles during surgery, signals from these natively innervated control sites can also be significantly increased. Additionally, subcutaneous fat can be removed as an adipofascial flap (as mentioned in Section 2.1.4 and described in Chapter 3) and used to physically separate target muscles (Figure 2.5).

2.2.3 Selective Transfer of Functionally Distinct Nerve Fascicles

Each of the large brachial plexus nerves contains axons that control a wide array of distal muscles. It is tempting to divide the nerve and transfer separate fascicles to different target muscles, in the hope that each target muscle would then produce unique EMG signals for different arm and hand functions. Unfortunately, there is little topographical organization in brachial plexus nerves at the proximal level; such organization only occurs more distally as nerves branch to the different muscles (Sunderland 1978; Jabaley et al. 1980; Stewart 2003). Thus separation of brachial plexus nerves and transfer of different fascicles to separate muscles might produce target muscles with different motor function profiles, but there is little chance that they would produce clean, distinct EMG signals for control of independent arm functions.

The one exception is the nerve to the triceps muscle in shoulder disarticulation patients. Through a series of cadaver dissections, we determined that the triceps nerve generally branches off the inferior posterior quadrant of the radial nerve, at about the glenohumeral fossa level. Thus, for shoulder disarticulation patients, the triceps nerve fascicle can be separated from the radial nerve for a short distance and transferred to a different motor nerve branch or muscle segment than the rest of the radial nerve. After reinnervation, the patient will then have two independent EMG signals: (1) the triceps nerve branch controlling elbow extension and (2) the distal radial nerve producing EMG signals for hand opening. We have successfully performed these transfers in shoulder disarticulation patients.

2.3 Conclusion

Successful TMR depends both on the inherent physiologic properties of severed nerves and target muscle, and on the physical parameters affecting EMG signal propagation and detection. To obtain robust reinnervation and to prevent control signal contamination by native EMG signals, target muscle must be entirely denervated. Subsequent hyper-reinnervation by large brachial plexus nerves ensures robust innervation of target muscle, even up to several years after amputation. Additionally, hyper-reinnervation may overcome the age-related reduction in peripheral nerve regeneration (Verdu et al. 2000; Ruijs et al. 2005): to date, TMR has been successfully performed in adults up to 55 years old. Animal studies indicate that hyper-reinnervation results in an increased number of motor units, suggesting that hyper-reinnervated muscle produces rich EMG signals that contain information intended for a large number of lost arm muscles. This is supported by high-density EMG studies in human TMR subjects that demonstrate different activation patterns corresponding to 16 intended movements of the missing hand, thumb, and fingers in reinnervated target muscles (Zhou et al. 2007). Finally, surgical removal of subcutaneous fat and separation of individual reinnervated muscle units are essential to achieve the strong, independent EMG signals necessary for conventional myoelectric control. Advances in understanding of the fundamental science behind nerve regeneration, muscle reinnervation, and EMG signal propagation will enable future modifications to surgical and signal processing techniques, to further enhance prosthetic control provided by TMR.

References

Davis, L. A., T. Gordon, J. A. Hoffer, J. Jhamandas, and R. B. Stein. 1978. Compound action potentials recorded from mammalian peripheral nerves following ligation or resuturing. *J Physiol* 285:543–559.

De la Barrera, E. J., and T. E. Milner. 1994. The effects of skinfold thickness on the selectivity of surface EMG. *Electroencephalogr Clin Neurophysiol* 93 (2):91–99.

Dhillon, G. S., S. M. Lawrence, D. T. Hutchinson, and K. W. Horch. 2004. Residual function in peripheral nerve stumps of amputees: implications for neural control of artificial limbs. *J Hand Surg [Am]* 29A (4):605–615.

Elsberg, C. A. 1917. Experiments on motor nerve regeneration and the direct neurotization of paralysed muscles by their own and by foreign nerves. *Science* 45 (1161):318–320.

Foehring, R. C., G. W. Sypert, and J. B. Munson. 1986. Properties of self-reinnervated motor units of medial gastrocnemius of cat. I. Long-term reinnervation. *J Neurophysiol* 55 (5):931–946.

Frey, M., H. Gruber, J. Hollie, and G. Freilinger. 1982. An experimental comparison of the different kinds of muscle reinnervation: nerve suture, nerve implantation, and muscular neurotization. *Plast Reconstr Surg* 69:656–667.

Fu, S. Y., and T. Gordon. 1995. Contributing factors to poor functional recovery after delayed nerve repair: prolonged axotomy. *J Neurosci* 15 (5 Pt 2):3876–3885.

Gordon, T., J. A. Hoffer, J. Jhamandas, and R. B. Stein. 1980. Long-term effects of axotomy on neural activity during cat locomotion. *J Physiol (Lond)* 303 (JUN):243–263.

Gordon, T., R. B. Stein, and C. K. Thomas. 1986. Innervation and function of hind-limb muscles in the cat after cross-union of the tibial and peroneal nerves. *J Physiol* 374:429–441.

Hoffman, H. 1950. Local re-innervation in partially denervated muscle—a histophysiological study. *Aust J Exp Biol Med Sci* 28 (4):383–397.

Jabaley M. E., W. H. Wallace, and F. R. Heckler. 1980. Internal topography of major nerves of the forearm and hand: a current view. *J Hand Surg [Am]* 5:1–18.

Jia, X. F., M. A. Koenig, X. W. Zhang, J. Zhang, T. Y. Chen, and Z. W. Chen. 2007. Residual motor signal in long-term human severed peripheral nerves and feasibility of neural signal-controlled artificial limb. *J Hand Surg [Am]* 32A (5):657–666.

Kuiken, T. A. 2003. Consideration of nerve-muscle grafts to improve the control of artificial arms. *J Tech Disabil* 15:105–111.

Kuiken, T. A., D. S. Childress, and W. Z. Rymer. 1995. The hyper-reinnervation of rat skeletal muscle. *Brain Res* 676 (1):113–123.

Kuiken. T. A., M. M. Lowery, and N. S. Stoykov. 2003. The effect of subcutaneous fat on myoelectric signal amplitude and cross-talk. *Prosthet Orthot Int* 27 (1):48–54.

Kuiken, T. A., G. A. Dumanian, R. D. Lipschutz, L. A. Miller, and K. A. Stubblefield. 2004. The use of targeted muscle reinnervation for improved myoelectric prosthesis control in a bilateral shoulder disarticulation amputee. *Prosthet Orthot Int* 28 (3):245–253.

Lowery, M. M., N. S. Stoykov, A. A. Taflove, and T. A. Kuiken. 2002. A multiple-layer finite-element model of the surface EMG signal, *IEEE Trans Biomed Eng* 49 (5):446–454.

Lowery, M. M., N. S. Stoykov, and T. A. Kuiken. 2003. Independence of myoelectric control signals examined using a surface EMG model. *IEEE Trans Biomed Eng* 50 (6):789–793.

Madison, R. D., G. A. Robinson, and S. R. Chadaram. 2007. The specificity of motor neuron regeneration (preferential reinnervation). *Acta Physiol* 189 (2):201–206.

O'Donovan, M. J., M. J. Pinter, R. P. Dum, and R. E. Burke. 1985. Kinesiological studies of self- and cross-reinnervated FDL and soleus muscles in freely moving cats. *J Neurophysiol* 54 (4):852–866.

O'Shaughnessy, K. D., G. A. Dumanian, R. D. Lipschutz, L. A. Miller, K. Stubblefield, and T. A. Kuiken. 2008. Targeted reinnervation to improve prosthesis control in transhumeral amputees. A report of three cases. *J Bone Joint Surg Am* 90 (2):393–400.

Ruijs, A. C., J. B. Jaquet, S. Kalmijn, H. Giele, and S. E. Hovius. 2005. Median and ulnar nerve injuries: a meta-analysis of predictors of motor and sensory recovery after modern microsurgical nerve repair. *Plast Reconstr Surg* 116 (2):484–494; discussion 495–496.

Scholz, T., A. Krichevsky, A. Sumarto, et al. 2009. Peripheral nerve injuries: an international survey of current treatments and future perspectives. *J Reconstr Microsurg* 25 (6):339–344.

Siemionow, M., and G. Brzezicki. 2009. Current techniques and concepts in peripheral nerve repair. In *Essays on Peripheral Nerve Repair and Regeneration*, edited by S. Geuna, P. Tos, and B. Battiston. San Diego: Elsevier Academic Press Inc.

Solomonow, M., R. Baratta, M. Bernardi, B. Zhou, Y. Lu, M. Zhu, and S. Acierno. 1994. Surface and wire EMG crosstalk in neighboring muscles. *J Electromyogr Kinesiol* 4 (3):131–142.

Stewart, J. D. 2003. Peripheral nerve fascicles: Anatomy and clinical relevance. *Muscle Nerve* 28 (5):525–541.

Sunderland, S. 1978. *Nerves and Nerve Injuries*. Edinburgh: Churchill Livingstone.

Thompson, W., and J. K. S. Jansen. 1977. Extent of sprouting of remaining motor units in partly denervated immature and adult rat soleus muscle. *Neuroscience* 2 (4):523–535.

Verdu, E., D. Ceballos, J. J. Vilches, and X. Navarro. 2000. Influence of aging on peripheral nerve function and regeneration. *J Peripher Nerv Syst* 5 (4):191–208.

Waller, A. 1850. Experiments on the sections of glossopharyngeal and hypoglossal nerves of the frog and observations of the alterations produced thereby in the structure of their primitive fibers. *Philos Trans R Soc Lond* 140:423–429.

Zhou, P., M. M. Lowery, K. B. Englehart, H. Huang, G. Li, L. Hargrove, J. P. Dewald, and T. A. Kuiken. 2007. Decoding a new neural machine interface for control of artificial limbs. *J Neurophysiol* 98 (5):2974–2982.

3

Surgical Techniques for Targeted Muscle Reinnervation

Gregory A. Dumanian and Jason M. Souza

CONTENTS

3.0 Introduction

The primary goal of targeted muscle reinnervation (TMR) surgery is to provide high-level upper limb amputees with independent myoelectric sites for prosthesis control. This translates to the preservation of existing sites and the creation of new sites through multiple nerve transfers. The TMR procedure was designed to create control sites for four basic prosthetic functions: elbow flexion, elbow extension, hand open, and hand close. However, because of the desirability of additional wrist and hand control and the potential benefits offered by advanced control algorithms, additional control sites should be created if possible. In this chapter, we present an overview of the surgical technique for transhumeral and shoulder disarticulation amputees, keeping in mind that the ultimate success of the procedure depends on the quality of the tissues of the residual limb, coordination between the surgical and rehabilitation teams, and the postsurgical rehabilitation process.

3.1 Fundamentals of Targeted Muscle Reinnervation Surgery

3.1.1 Brachial Plexus Anatomy and Innervation

The TMR procedure is based on an understanding of the anatomy and function of four major infraclavicular branches of the brachial plexus (Figure 3.1). Because this information guides both the surgical procedure and the subsequent approach to prosthetic treatment, it is reviewed here briefly, with a summary presented in Table 3.1.

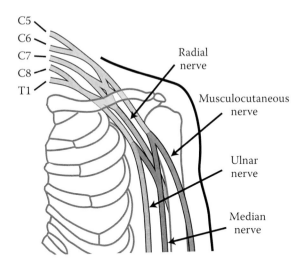

FIGURE 3.1
(See color insert.) Diagram of brachial plexus anatomy.

TABLE 3.1

Summary of Arm Movements Controlled by Brachial Plexus Nerves

	Arm Movement							
Nerve	Elbow Flexion	Elbow Extension	Hand Open	Hand Close	Wrist Pronation	Wrist Supination	Wrist Flexion	Wrist Extension
Musculocutaneus nerve	X							
Radial nerve		X	X			X		X
Median nerve				X	X		X	
Ulnar nerve			X	X			X	

3.1.1.1 Musculocutaneous Nerve

The musculocutaneous nerve arises from the lateral cord of the brachial plexus. It normally innervates the coracobrachialis muscle, which provides shoulder flexion and adduction; the biceps brachii, which provides elbow flexion and forearm supination; and the brachialis, which supplies isolated elbow flexion. Damage to the musculocutaneous nerve causes a marked deficiency in elbow flexion.

3.1.1.2 Radial Nerve

The radial nerve originates from the posterior cord of the brachial plexus and is the largest in diameter of the brachial plexus nerves. In the upper arm, it innervates the triceps brachii, anconeus, and brachioradialis, which act to extend, stabilize, and flex the elbow, respectively. In the forearm,

the radial nerve innervates the extensor carpi radialis longus and bre-
vis, which extend and radially deviate the wrist. The radial nerve then
branches into the posterior interosseus and superficial radial nerves, the
former of which serves to innervate the supinator and extrinsic wrist and
finger extensors.

3.1.1.3 Median Nerve

The median nerve is derived from both medial and lateral cords of the brachial plexus. In the forearm, its branches innervate the superficial, interme-
diate, and deep flexor muscles, which primarily act to pronate the forearm
and flex the wrist, fingers, and thumb. In the hand, it innervates the thenar
muscles, which flex and abduct the thumb, and the first and second lumbri-
cals, which flex the metacarpophalangeal joints and extend the interphalangeal (IP) joints of the first and second digits.

3.1.1.4 Ulnar Nerve

In the forearm, the ulnar nerve supplies the flexor carpi ulnaris and the
medial half of the flexor digitorum profundus, which act to flex the wrist
and fingers, respectively. At the wrist, the ulnar nerve divides into deep and
superficial branches. The deep branch innervates intrinsic hand muscles that
flex, abduct, and adduct the fingers and extend the IP joints, while the superficial branch is predominantly sensory in function.

3.1.2 Nerve Selection for Targeted Muscle Reinnervation

The nerve transfers performed during the TMR procedure are designed to
regain the motor control information that is still present in the peripheral
nervous system but has become inaccessible because of loss of muscle effectors by amputation. After successful nerve transfers, the musculocutaneous,
median, radial, and ulnar nerves cause contraction of the target muscles to
which they have been coapted. The electromyographic (EMG) signals from
these muscles are used to control one of numerous functions in the prosthetic limb, based on the needs of the patient and the requirements of the
prosthetist (see Table 3.1).

3.1.3 Patient Selection

The minimum eligibility requirements for TMR surgery include an upper
extremity amputation and a desire for improved prosthetic function.
Patients with spinal cord or proximal brachial plexus injuries should not
undergo the procedure. TMR was developed for individuals with shoulder disarticulation or transhumeral amputations. The transhumeral TMR
procedure requires residual biceps and/or triceps muscles that are under

active control and a residual limb that is long enough to operate a prosthetic arm effectively. Very short transhumeral amputations or those without cortically controlled biceps or triceps muscles are best treated as shoulder disarticulations. TMR for below-elbow amputations is theoretically possible but would require significant advancements in prosthetic limb design and manufacture.

Each TMR candidate is evaluated as an individual, with an effort to synthesize the physical characteristics of the residual soft tissues and nerve function with patient expectations, understanding, and anticipated compliance with rehabilitation. Candidates should have a thorough understanding of the surgical process and risks and be motivated to participate in postsurgical rehabilitation and training. The time interval since amputation does not seem to influence the outcome of the TMR procedure; however, nerve coaptations in younger patients may exhibit an improved ability to produce functional EMG signals due to the increased capacity for nerve regeneration in younger individuals (Verdú et al. 2000). Supple soft tissues and an absence of previous flaps or skin grafts to the residual limb facilitate the surgical procedure, although prior surgery to the soft tissues of the residual limb is by no means a contraindication to TMR. Residual limb length is directly related to the difficulty of the TMR procedure: long nerves permit more numerous nerve transfers and allow the surgeon to excise more marginal nerve tissue adjacent to the end-neuromas of the donor nerves. In the absence of available actively controlled muscle in the vicinity of the donor nerves, free tissue transfer will be required to provide a suitable target muscle. Surgical risks include bleeding, infection, paresthesias, loss of sensation, worsening of phantom pain, and the need for additional surgical procedures. As with any nerve transfer procedure, there is always the possibility that the TMR procedure will fail to provide any functional benefits.

Following the procedure, patients should refrain from wearing their prostheses for a period of 3 to 6 weeks to allow for adequate healing of the surgical site. Once healed, the patients are then refitted with a standard prosthesis. Prior to the procedure, patients must be made aware that it generally takes 5 to 7 months for the target muscles to become fully reinnervated. Thus, they should be counseled to expect this delay, as final fitting of the TMR-controlled prosthesis cannot proceed until reinnervation has occurred. Patients should also be informed of the potential for the skin overlying the reinnervated muscles to be reinnervated by afferent fibers of the transferred nerves. This process can produce the perception of touch on the missing limb when touch is applied to the reinnervated skin (see Chapter 8).

3.1.4 Physical Examination

Patients must undergo a thorough evaluation, including a review of their medical history and a physical exam, before being considered candidates for surgery. The details of the amputation provide great insight into whether any avulsive injury was sustained at the time of amputation, as this can dictate the location of the end-neuromas in the residual limb. The operative report from the amputation surgeon will often provide useful details regarding previous management of residual limb nerves. The degree of distal nerve damage can also be inferred from any available photographs taken at the time of the initial presentation after traumatic injury. Surgical amputation is commonly combined with *traction neurectomies*, whereby the amputated nerves are placed under tension and then transected to allow the cut nerves to retract into unscarred soft tissues. The cut ends of these large mixed nerves inevitably form end-neuromas, and the more proximal these end-neuromas are, the more difficult they are to locate and mobilize during surgery. During a physical examination, gentle tapping on the soft tissues can often elicit a Tinel's sign, signifying the location of the end-neuroma of a major peripheral nerve. Feedback from the patient can indicate whether the median, ulnar, radial, or musculocutaneous nerve is being stimulated. It is also important to document which muscles are under active cortical control. The presence of local muscles that are palpable but are atrophied and lack cortical control may suggest a proximal brachial plexus injury, which would preclude performance of the TMR procedure. Patient evaluation is largely clinical, with little information provided by magnetic resonance imaging (MRI) or electromyography (EMG). However, imaging with plain films can be useful for documenting the length of the residual bone and relative soft tissue envelope, and for comparison with plain films of the contralateral extremity. The presence of a small length of humerus in a shoulder disarticulation patient implies that the nerves will probably be of sufficient length to allow for nerve transfer. In some cases, the residual humerus may not be long enough to control the prosthesis in space, but there may be enough remaining native biceps and triceps to enable control of the prosthetic elbow without the need for nerve transfers. Unstable soft tissue envelope and open wounds are additional relative contraindications for surgery but may be amenable to coverage by tissue expansion or flap procedures prior to TMR. Finally, heterotopic ossification, which can be identified radiographically, has been observed to encase end-neuromas in bone, rendering their locations quite difficult to ascertain by physical exam.

3.2 Targeted Muscle Reinnervation for the Transhumeral Amputee

3.2.1 Surgical Goals

TMR for the transhumeral amputee comprises two similar procedures performed on the ventral and dorsal aspects of the residual upper arm

(a) Key muscles for transhumeral TMR

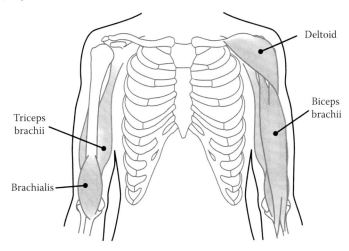

(b) TMR for transhumeral amputation

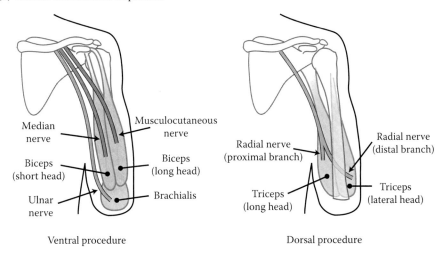

Ventral procedure   Dorsal procedure

FIGURE 3.2
(See color insert.) (a) Normal anatomy of key muscles for transhumeral TMR surgery. (b) Schematic of typical surgical plan for transhumeral TMR.

(O'Shaughnessy et al. 2008; Dumanian et al. 2009) (Figure 3.2). On the ventral side, the innervation of the long head of the biceps muscle is left intact to provide an intuitive elbow flexion signal. The median nerve is transferred to the motor nerve innervating the short head of the biceps to provide a "hand close" signal. On the dorsal side, the native innervation to the long head of the triceps by the proximal radial nerve is left intact to provide an elbow extension signal. The distal radial nerve is transferred to the motor nerve of the lateral head of the triceps to provide a "hand open" signal. If the residual limb is long enough, the ulnar nerve is transferred to the motor nerve of the brachialis muscle to provide additional hand or wrist control signals.

In the process of exposing the nerves and target muscles on each side of the residual arm, proximally based adipofascial flaps are created; this procedure thins the overlying skin flaps and provides tissue to be used as a spacer between target muscle segments. This manipulation of the overlying soft tissue allows for strong, well-differentiated EMG signals to be recorded from each target muscle. A 20-minute training video that provides an overview of TMR and the surgical procedure for a transhumeral amputee (including patient selection, technique, and rehabilitation) is available on the Center for Bionic Medicine website.

3.2.2 Surgical Procedure

Prior to surgery, the borders of the biceps and triceps muscles are confirmed by having the patient attempt to flex and extend the missing elbow. A line is drawn representing the putative median raphe between the short and long heads of the biceps and the long and lateral heads of the triceps. The surgical procedure is performed with general anesthesia but without the use of muscle relaxers in order to allow for effective nerve stimulation.

3.2.2.1 Ventral Dissection

The patient is placed in a supine position with the residual arm abducted and placed on an arm board for support. Ventral exposure is gained through an incision made directly over the belly of the biceps muscle, beginning just inferior to the deltoid and running distally between the two heads of the muscle. Dilute epinephrine solution (1:500,000) may be used before and after the incision is made to improve hemostasis and aid visualization of the tissue planes. However, this might interfere with effective use of a nerve stimulator and should be limited to the skin and superficial subcutaneous tissue. Thin skin flaps are elevated on either side of the incision for approximately 2 to 3 cm. A proximally based, U-shaped adipofascial flap is then raised (see Figure 3.3). The brachial fascia overlying the biceps is included in the flap. Following elevation of the adipofascial flap, the interspace between the heads of the biceps is further developed. The surface of the biceps is carefully evaluated for the thin raphe that will permit a bloodless approach to the musculocutaneous nerve. The dissection continues between the muscles to expose the musculocutaneous nerve, the motor branches to the short and long heads of the biceps, and the motor nerve to the brachialis muscle (Figure 3.4). The native innervation of the long head of the biceps by the musculocutaneous nerve is left intact to preserve the elbow flexion signal. On occasion, there

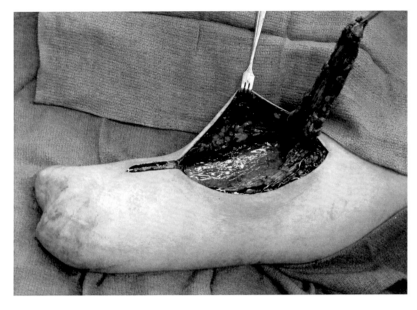

FIGURE 3.3
(See color insert.) An adipofascial flap is raised during ventral dissection of the residual limb in a transhumeral amputee. Following nerve coaptation, the flap is placed between the long and short heads of the biceps brachii to facilitate isolation of the two EMG signals.

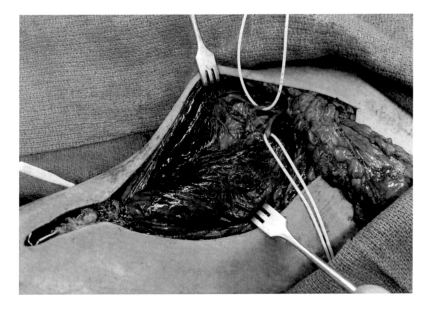

FIGURE 3.4
(See color insert.) Ventral exposure revealing the motor branches of the musculocutaneous nerve to the long and short heads of the biceps.

is more than one motor branch that innervates the long or short head of the biceps. These accessory branches should be identified, and then transected or preserved based on the goals of the procedure.

3.2.2.2 Median Nerve Transfer

The median nerve can be identified by carefully lifting the belly of the short head of the biceps away from the humerus. Following transhumeral amputation, the median nerve is no longer contiguous with its target musculature; thus its identity can only be determined by anatomic position and size. A mid-size sensory nerve adjacent to the brachial artery, which eventually becomes the medial antebrachial cutaneous nerve, is typically present and is usually the first nerve identified while searching for the median nerve. The neuromatous end of the median nerve is resected back to healthy-looking fascicles in order to maximize its regenerative capacity. The motor nerve to the short head of the biceps (a branch of the musculocutaneous nerve) is then divided as close to the muscular entry point as possible in an effort to minimize reinnervation time. The median nerve is then coapted directly to the motor nerve(s) of the short head with loop magnification and 5-0 or 6-0 Prolene™ sutures (Ethicon, Somerville, NJ). A suture is used to deliver the small motor nerve into the center of the larger median nerve stump. A second suture can be used to secure the coaptation construct to surrounding epimysium for additional stability.

3.2.2.3 Ulnar Nerve Transfer

A long residual arm, produced by amputation at or near the level of the epicondyles, will often be amenable to an additional ventral nerve transfer. Based on its location posterior to the brachial artery, the ulnar nerve is unlikely to be the first major nerve visualized from this ventral exposure. The ulnar nerve is identified posterior to the intermuscular septum, and it is typically smaller than the median nerve. The motor nerve to the brachialis is identified as the major distal branch of the musculocutaneous nerve, which arises after the branching off of the motor nerves to the biceps muscle. The ulnar nerve is mobilized, cut back to healthy fascicles, and delivered through the biceps muscle to rest adjacent to the motor branch to the brachialis. The nerve coaptation is performed with an analogous technique to that used with the median nerve. If the overlying biceps muscle limits the extent to which the brachialis is expected to produce a recordable surface EMG signal, 1 to 2 cm along the lateral edge of the distal biceps can be excised to make the brachialis more superficial and produce an EMG signal that is distinct from that produced by the long head of the biceps.

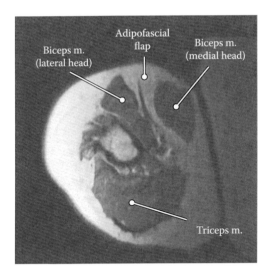

FIGURE 3.5
MRI showing the adipofascial flap placed between the lateral (*long*) and medial (*short*) heads of the biceps muscle to ensure independent EMG signal recordings. (Reprinted from O'Shaughnessy et al., Targeted reinnervation to improve prosthesis control in transhumeral amputees. A report of three cases, *J Bone Joint Surg* 90 (2):393–400, 2008. With permission.)

3.2.2.4 Completion of Ventral Procedure

Following nerve coaptation, the previously raised adipofascial flap is placed between the short and long heads of the biceps. The adipofascial flap creates space between the two heads of the biceps muscle to assist with eventual separation of EMG signals (Figure 3.5). The flap also acts as a barrier to prevent aberrant nerve growth between the cut nerve endings of the musculocutaneous nerve and the coaptation sites. The skin is closed with either sutures or staples, and a drain or compressive dressing is used to eliminate the dead space created by the dissection.

3.2.2.5 Dorsal Procedure

While the dorsal procedure can be accomplished with the patient supine and the shoulder placed in an extended position, the radial nerve dissection is much easier with the patient in the prone position. An incision is made on the residual limb over the presumed course of the raphe between the long and lateral heads of the triceps. The interdigitation between the muscles is variable—the space is more easily found just inferior to the deltoid. The triceps is split between the long and lateral heads to reach the radial nerve as it courses adjacent to the humerus in the spiral groove. Within this space, the radial nerve and its motor branch to the lateral head of the triceps are revealed (Figure 3.6). The native innervation of the long

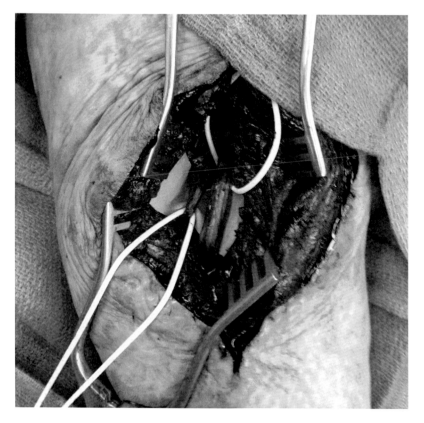

FIGURE 3.6
(See color insert.) Dorsal exposure revealing the radial nerve and its motor branch to the lateral
head of the triceps muscle.

head of the triceps by the proximal radial nerve is left intact to preserve the
elbow extension signal. As the motor nerves to the long head are relatively
proximal to the motor nerves of the lateral head, they are often not visualized with this approach. The distal radial nerve is identified and cut back to
healthy-looking fascicles. The motor nerve to the lateral head of the triceps
(a branch of the proximal radial nerve) is cut near its entry point into the
muscle. The distal radial nerve is transferred in a tension-free manner to
the cut motor nerve of the lateral head of the triceps, similarly to the coaptation of the median nerve on the ventral side. Variations to this procedure
are possible, based on the anatomy of the local motor nerves. Overall, the
goal is to transfer the healthy distal radial nerve to a functional segment of
triceps, while preserving the innervation to the remaining triceps. Drains
and a mildly compressive dressing are then applied, and the skin is closed
in a similar fashion to that of the ventral closure.

3.2.3 Options and Special Considerations

3.2.3.1 Limiting Target Muscle Mobility

The loss of distal insertion points for the biceps muscle in short transhumeral amputations may cause the muscle to pull out of the prosthetic socket and under the deltoid muscle when contracting. This makes it difficult to capture an EMG signal from this muscle using socket-mounted electrodes. To allow for stable EMG recording, the muscle may be disoriginated to prevent it from migrating upward during contraction.

3.2.3.2 Angulation Osteotomy

For a transhumeral amputee with a long residual humerus, TMR can be combined with a wedge osteotomy to produce an anteriorly angulated residual limb. This technique has been refined by use of a dorsal reconstruction plate to yield predictable and durable angulation of the residual humerus (Figure 3.7). The anterior angulation produced by this technique provides the prosthetist with an additional point of suspension and translates into better rotational control of the prosthesis by the patient.

3.3 Targeted Muscle Reinnervation for the Shoulder Disarticulation Amputee

TMR for shoulder disarticulation amputees is more challenging than for transhumeral amputees; this is largely attributed to greater anatomic distortion and shorter donor nerve length. Although the overall surgical strategy can be outlined ahead of time, each procedure must be individually tailored to the nervous and muscular anatomy observed intraoperatively. Frequently, intraoperative findings demand divergence from the preoperative plan, requiring the surgical team to be flexible and to have expertise in upper chest and axillary anatomy.

3.3.1 Surgical Goals

3.3.1.1 Overview

TMR for the shoulder disarticulation patient is designed to create independent myoelectric control sites for four basic prosthesis functions: hand open, hand close, elbow flexion, and elbow extension (Hijjawi et al. 2006). Additional sites are desirable and should be created if possible. This is accomplished by first identifying candidate target muscles and their motor nerve innervation, identifying the amputated nerves as they emerge off of

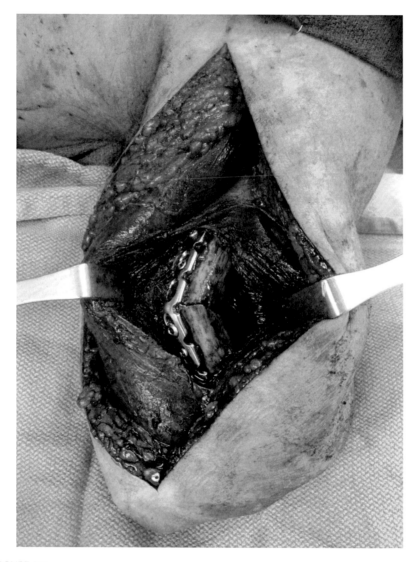

FIGURE 3.7
(a) (See color insert.) Anterior view of angulation osteotomy in a transhumeral amputee.
(continued)

the brachial plexus, denervating the target muscles, performing the nerve transfers, and manipulating the overlying soft tissues to improve later EMG signal detection.

3.3.1.2 Potential Target Muscles

Potential target muscles include the pectoralis major, pectoralis minor, serratus anterior, and latissimus muscles (Figure 3.8). From a theoretical

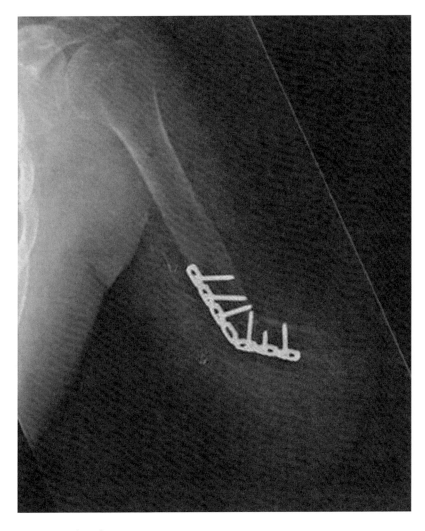

FIGURE 3.7 (continued)
(b) X-ray of angulation osteotomy in a transhumeral amputee.

standpoint, the number of target segments into which the pectoralis major can be divided is limited only by the number of motor entry points. However, in no previous TMR surgery has the pectoralis been divided into more than three segments, and prior experience suggests that it is most practical to divide the pectoralis major into a clavicular segment and a split sternal segment. The plane between the clavicular and sternal heads of the pectoralis major is most easily identified toward the insertion of the muscle. Once this plane is entered, the two segments are easily separated from each other via blunt dissection (Figure 3.9). In several patients, the sternal head has subsequently been divided into two target segments and

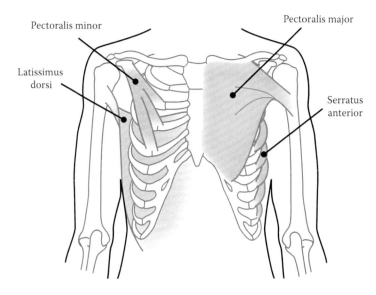

Pectoralis minor

Pectoralis major

Latissimus
dorsi

Serratus
anterior

FIGURE 3.8
(See color insert.) Normal anatomy of key musculature for shoulder disarticulation TMR
surgery.

FIGURE 3.9
(See color insert.) Plane between the clavicular (left) and sternal (right) heads of the pectoralis
major muscle. The brachial plexus lies within the fatty tissue exposed by this dissection.

has successfully yielded two distinct EMG signals in all cases. The pectoral nerves enter the deep surface of the muscle after emerging through and around the pectoralis minor muscle. Based on the course and branching pattern of the individual motor branches of the pectoral nerves, the pectoralis major muscle is then split along its muscle fibers to produce two independent target muscle segments. The viability of the separated segments of the sternal head is maintained via blood supply from the internal mammary artery. In the case of a female patient, it is desirable to keep signal detection sites high on the chest away from the thicker soft tissues of the breast; thus the clavicular head may be split to provide an alternative to the sternal head. This requires that the clavicular head exhibit two distinct motor entry points. The pectoralis minor can be used as an additional target site by mobilizing it laterally off of the chest wall. Likewise, the thoracodorsal or long thoracic nerves can be used as recipients for nerve transfer to the latissimus or serratus muscles, respectively. Once the patient's anatomy has been delineated intraoperatively, the final pattern of nerve transfers can be mapped out, with the goal of providing the optimal combination of EMG recording sites. The intraoperative appearance of the donor nerves should be considered alongside expected prosthetic functions to ensure that the most important functions are paired with nerve transfers that are most likely to produce robust, accessible EMG signals. The musculocutaneous nerve, solely responsible for elbow flexion, should be transferred to an optimally located muscle segment. The hierarchy of donor nerves continues with the median nerve, the radial nerve, and lastly the ulnar nerve. The ulnar nerve is given lowest priority as a result of its diverse pattern of motor innervation within the hand and forearm, which can lead to an unpredictable pattern of reinnervation. This hierarchy achieves prosthetic function in order of importance with the key functions being elbow flexion, followed by hand close, elbow extension, and finally hand open. If available, additional sites can be paired to wrist pronation or supination, but any attempts to provide wrist function should not compromise the primary functions previously outlined.

3.3.1.3 Typical Surgical Plan

A typical surgical plan is as follows: The musculocutaneous nerve is transferred to the clavicular head of the pectoralis major to provide the elbow flexion signal. The lack of overlying adipose tissue and the ability of the clavicle to conduct electrical currents combine to increase the EMG amplitude and make this an optimal location for signal recording (Lowery et al. 2002). The median nerve is most often transferred to the largest motor nerve innervating the sternal head of the pectoralis major and is intended to provide the hand close signal. In some patients, a proximal branch of the radial nerve remains connected to the residual triceps. If this is the case, this connection should be left intact to preserve the elbow extension signal. The distal branch of the radial nerve can then be transferred to a remaining target muscle site—often the latissimus muscle via coaptation to the thoracodorsal nerve—to provide the hand open signal. If the proximal branch of the radial nerve is no longer connected to the triceps, the entire radial nerve can be transferred to the thoracodorsal or long thoracic nerves. Alternatively, the radial nerve has been successfully transferred to a pectoral nerve innervating the inferolateral border of the pectoralis major muscle, although this requires considerable residual radial nerve length. This pectoral motor nerve has also served as the recipient for a coaptation to the ulnar nerve. As previously mentioned, the pectoralis minor can act as an additional nerve transfer target, although it must be mobilized laterally into the axilla to provide a surface recording location that is distinct from the pectoralis major sites.

3.3.2 Surgical Procedure

Just as with the transhumeral procedure, surgery is done under general anesthesia without the use of muscle relaxers. To limit additional scarring, the previous amputation incisions can be reopened and deepened to provide access to the brachial plexus and target muscles. Optimal exposure is provided by an incision over the interspace between the sternal and clavicular heads of the pectoralis major, since the majority of the dissection takes place in this space.

3.3.2.1 *Expose Pectoralis Muscles and Identify Innervation*

Skin flaps are elevated to expose the pectoralis major muscle. Excessive fat over the target muscles can be thinned at this point to improve future EMG signal detection, or mobilized as a medially based flap for later placement between muscle segments. Two dissection planes are developed: one between the sternal and clavicular heads of the pectoralis major muscle and the other between the pectoralis major and minor muscles. The motor nerves to each muscle segment are identified and tagged at their entry points. The motor nerves to the clavicular head of the pectoralis major arise from the lateral cord of the brachial plexus and can be located at the midpoint of the clavicle. They travel from deep to superficial and can be visualized through the interspace between the sternal and clavicular muscle segments. The motor nerve to the clavicular head is accompanied by a vascular pedicle, requiring meticulous tissue handling to prevent bleeding. Motor nerves to the sternal head come in three clusters: one is located near the motor nerve to the clavicular head, one travels through or is just medial to pectoralis minor, and one emerges lateral to pectoralis minor and enters the inferolateral aspect of pectoralis major.

3.3.2.2 Identify Brachial Plexus

To locate the brachial plexus and its emerging branches, dissection then proceeds laterally to the pectoralis major, near the insertion of the pectoralis minor tendon. The musculocutaneous, median, radial, and ulnar nerves are identified by their branching patterns and relative positions and resected back to healthy-looking fascicles. The radial nerve is stimulated to determine if any remnant triceps innervation is left intact. If so, this branch is preserved and any scar tissue and excess fat overlying the remnant triceps is excised to improve later EMG signal detection. If there is no remnant triceps innervation, the radial nerve should be carefully examined to determine if it can be split into its posterior and anterior aspects, which formerly innervated the elbow extensors and wrist/hand extensors, respectively. If additional donor nerve length is required, dissection should proceed proximally along the brachial plexus, after which the donor nerves can be mobilized medially around the pectoralis minor tendon. This allows the donor nerves to be routed directly to the anterior chest rather than coursing through the axilla.

3.3.2.3 Expose Motor Nerves to Target Muscles

As previously described, the pectoralis major can be divided into several separate muscle regions. Having already separated the clavicular head from the sternal head, the sternal head is split into upper and lower segments based on the nerve branching and vascular anatomy. Prior to performing the targeted nerve transfers, the pectoralis major muscle must be completely denervated to prevent native motor innervation from the pectoral nerves from interfering with reinnervation by the donor nerves. The deep surface of the pectoralis major should be thoroughly exposed to ensure that no remnant motor nerves have been left intact. This is particularly important medially, where exposure is likely to be most limited.

Target muscle segments should be at least 1 cm thick and 3 to 5 cm wide to ensure sufficient EMG signal strength. If the pectoralis minor is to be used as a transfer target, any accessory native motor nerves to the pectoralis minor should be identified and divided prior to the nerve transfer. The pectoralis minor is then mobilized as an island flap based on its lateral neurovascular pedicle. Alternatively, the pectoralis minor can be repositioned into the axilla as a turnover flap, rotating it laterally on its long axis while leaving its lateral border intact. Either way, care must be taken to preserve its vascular supply. Should an additional or alternative muscle target be needed, the posterior (spinal) deltoid can be used. However, technical challenges associated with exposure and coaptation to the axillary nerve restrict the use of the deltoid as a primary muscle target.

3.3.2.4 Identify and Dissect Thoracodorsal and/or Long Thoracic Nerve

If the latissimus and/or serratus anterior muscles have been selected as target muscles, the thoracodorsal (for latissimus) or long thoracic (for serratus anterior) nerves must be identified and dissected. With this approach, the thoracodorsal nerve is identified deep to the plexus as it emerges from the posterior cord of the brachial plexus and travels deep to the lateral border of the latissimus muscle. The long thoracic nerve travels anteriorly in relation to the thoracodorsal nerve, running along the chest wall on the superficial surface of the serratus anterior muscle. A nerve stimulator is useful in the identification of these nerves. The points of motor nerve entry into the serratus muscle are relatively distal compared to other potential target muscles. Thus use of the long thoracic nerve as a recipient is usually limited to the rare patient who possesses long donor nerves but insufficient biceps or triceps to allow for nerve transfer based on the transhumeral pattern of reinnervation.

3.3.2.5 Perform Nerve Transfers

Once the target muscles have been identified and isolated, and their native innervations have been disrupted, the nerves transfers are performed. End-to-end nerve coaptations are performed as close as possible to the motor entry point into the muscle. This reduces the amount of time needed for reinnervation. If possible, the nerve coaptations are separated from each other spatially to decrease the chance for aberrant innervation.

3.3.2.6 Isolate Muscle Segments and Close Skin Flaps

To isolate the target muscle segments, adipofascial flaps or free fat grafts are positioned between the pectoralis major muscle segments (Figure 3.10). The adipose tissue overlying the muscle segments is directly thinned where possible. In the case of the latissimus or serratus muscles, excess adipose tissue can be removed with liposuction techniques. If remnants of biceps or triceps remain, these should be oriented to allow for maximal isolation, and the overlying adipose tissue should be removed. The skin flaps are then closed over drains. Quilting sutures placed percutaneously between the skin and the chest wall have been used to minimize seroma formation.

3.3.3 Options and Special Considerations

3.3.3.1 Free Tissue Transfer

Free tissue transfer has been used to create a muscular target when no cortically controlled muscle is locally available. Free tissue transfer offers the potential to fill a soft tissue deficiency while providing muscle for TMR. As previously reported, a free serratus muscle flap can be used to cover a

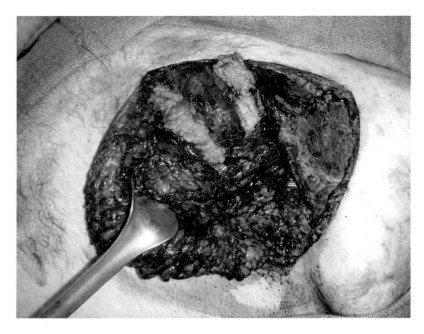

FIGURE 3.10
(See color insert.) Medially based adipofascial flaps positioned between the clavicular head and two sternal segments of the pectoralis major muscle.

shoulder disarticulation site, with each slip of the serratus serving as a separate target for reinnervation by different donor nerves (Figure 3.11) (Bueno et al. 2011). Other muscle flaps can be used in a similar fashion, with the potential for reinnervation dictated by the pattern of innervation. The rectus abdominis muscle, for example, has been shown to have a segmental pattern of innervation, with at least three separate compartments, each innervated by a distinct thoracoabdominal nerve (Rozen et al. 2008). For the purposes of EMG recording, skin grafting is an ideal method for muscle flap coverage because it minimizes the barrier between the surface electrode and the target muscle.

3.3.3.2 Considerations for Female Shoulder Disarticulation Patients

As stated previously, it is often desirable to perform coaptations to target muscles higher on the chest of female patients, so as to limit the need to thin breast tissue over target muscles. The upper pole of the breast is frequently thin enough to allow for use of the upper portion of the sternal head of the pectoralis muscle, but use of the lower portion of the sternal head is almost always prohibited by the overlying breast tissue. It is therefore desirable to split the clavicular head of the pectoralis major to create two independent

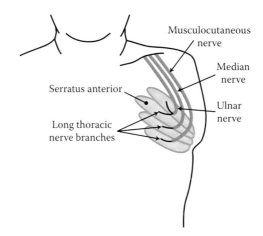

FIGURE 3.11
Free serratus muscle flap transfer with split-thickness skin graft. The individual slips of the serratus muscle were reinnervated by donor brachial nerves according to the pattern illustrated. (Modified from Bueno et al., Targeted muscle reinnervation in a muscle-free flap for improved prosthetic control in a shoulder amputee: case report, *J Hand Surg [Am]* 36A (5):890–893. © [2011]. With permission from Elsevier.)

myoelectric control sites. If the clavicular head lacks multiple motor nerve entry points, the serratus muscle can be used as an alternative target.

3.4 Targeted Sensory Reinnervation

An unexpected result of the first targeted reinnervation surgery, which was performed on an individual with a shoulder disarticulation amputation, was that the skin overlying the target muscles appeared to be reinnervated by the sensory afferents of the transferred nerves. This skin had been denervated during surgery by the removal of subcutaneous tissue. Stimulation of this reinnervated skin produced the perception that the missing hand and arm were being stimulated.

While sensory transfer in shoulder disarticulation amputees has occurred spontaneously following targeted muscle reinnervation, it does not occur predictably unless a region of skin overlying the target muscles is completely denervated. Likewise, in transhumeral amputees, spontaneous sensory reinnervation by donor nerves has only been demonstrated following complete denervation of an area of residual limb skin, for example, the reinnervation of the medial arm skin overlying the biceps in cases where the medial antebrachial cutaneous nerve was transected. In general, however, transhumeral patients do not frequently exhibit clinically significant denervation of the upper arm skin following the TMR procedure, and, therefore, spontaneous sensory reinnervation following TMR for transhumeral amputees is rare.

Modifications to the TMR procedure have been developed in an attempt to foster and direct the process of sensory reinnervation. Given its proximity to the site of the motor nerve transfers, the supraclavicular nerve has been utilized as a sensory nerve recipient. After the proximal nerve stump has been sufficiently buried to prevent auto-reinnervation, the distal stump of the supraclavicular nerve is coapted to the median nerve in an end-to-side fashion with the aim of establishing a sensory representation of the missing hand over the region of skin previously supplied by the supraclavicular nerve.

This sensory transfer provides a potentially exciting portal to provide an amputee with anatomically appropriate sensory feedback from the prosthesis. This has been achieved in the laboratory setting, but there is not yet a viable way to provide this feedback to a patient clinically; thus there is no benefit to the patient at this time. For more details pertaining to targeted sensory reinnervation, see Chapter 8.

References

Bueno, R. A., B. French, D. Cooney, and M. W. Neumeister. 2011. Targeted muscle reinnervation of a muscle-free flap for improved prosthetic control in a shoulder amputee: case report. *J Hand Surg [Am]* 36A (5):890–893.

Dumanian, G. A., J. H. Ko, K. D. O'Shaughnessy, P. S. Kim, C. J. Wilson, and T. A. Kuiken. 2009. Targeted reinnervation for transhumeral amputees: current surgical technique and update on results. *Plast Reconstr Surg* 124 (3):863–869.

Hijjawi, J. B., T. A. Kuiken, R. D. Lipschutz, L. A. Miller, K. A. Stubblefield, and G. A. Dumanian. 2006. Improved myoelectric prosthesis control accomplished using multiple nerve transfers. *Plast Reconstr Surg* 118 (7):1573–1578.

Lowery, M. M., N. S. Stoykov, A. Taflove, and T. A. Kuiken. 2002. A multiple-layer finite-element model of the surface EMG signal. *IEEE Trans Biomed Eng* 49 (5):446–454.

O'Shaughnessy, K. D., G. A. Dumanian, R. D. Lipschutz, L. A. Miller, K. Stubblefield, and T. A. Kuiken. 2008. Targeted reinnervation to improve prosthesis control in transhumeral amputees. A report of three cases. *J Bone Joint Surg* 90 (2):393–400.

Rozen, W. M., M. W. Ashton, B. J. Kiil, D. Grinsell, S. Seneviratne, R. J. Corlett, and G. I. Taylor. 2008. Avoiding denervation of rectus abdominis in DIEP Flap Harvest II: an intraoperative assessment of the nerves to rectus. *Plastic Reconstruct Surg* 122 (5):1321–1325.

Verdú, E., D. Ceballos, J. J. Vilches, and X. Navarro. 2000. Influence of aging on peripheral nerve function and regeneration. *J Peripher Nerv Syst* 5 (4):191–208.

4

Targeted Muscle Reinnervation as a Strategy for Neuroma Prevention

Jason H. Ko, Peter S. Kim, and Douglas G. Smith

CONTENTS

4.0 Introduction

When a peripheral nerve is transected, a bulbous swelling, or *neuroma*, will develop at the severed end of the nerve (Figure 4.1). Neuromas consist of disorganized, chaotic axons encased in significant scar and fibrous tissue. They are frequently sensitive to pressure and other forms of mechanical irritation, causing a focal pain that is often difficult to treat. By providing a distal muscle target for the proximal transected axons to grow into, targeted muscle reinnervation (TMR) may represent a novel technique for the prevention and treatment of neuromas. In this chapter, we will discuss the pathophysiology

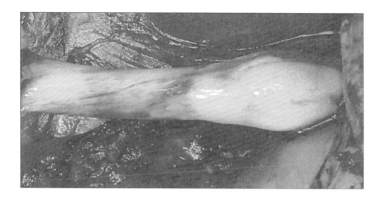

FIGURE 4.1
(See color insert.) Median nerve end-neuroma encountered during a transhumeral TMR procedure.

of neuromas, describe existing treatments, and outline the potential role that TMR may play in their prevention and treatment.

4.1 Neuroma Pathophysiology

Early descriptions of painful neuromas date back to 1634, when Ambroise Paré described the treatment of neuromas with oils and massage (Paré, 1634). It was not until 1811 that Louis Odier delineated an anatomic structure that was responsible for painful sensations long after injury, and stated that the bulbous stump of a transected nerve was sensitive (Odier, 1811). Following Odier's discovery, in 1828, William Wood performed the first pathologic studies of cut nerve endings and coined the term *neuroma* (Wood, 1828). These studies were further elaborated on by Rudolf Virchow (1863). Although our understanding of the peripheral nervous system and the responses of its component parts to injury has grown immensely in the 200 years since Odier first described the painful neuroma, many questions remain unanswered.

4.1.1 Etiology

When a peripheral nerve is transected, the distal nerve segment undergoes a process first described by Waller in 1850 and now accordingly known as Wallerian degeneration (Waller, 1850). The distal nerve segments start to degenerate and, if connection to the proximal nerve segment is not reestablished, the distal segment eventually disappears. In addition to describing changes in the distal nerve segment, Waller also described the generation of neural tissue from the proximal nerve; this process was further characterized by Ramón y Cajal in 1905. Within 1 to 2 days of injury, multiple unmyelinated axon sprouts grow from the cut proximal ends of axons. If these regenerating axon sprouts fail to reach a distal nerve target, enter distal Schwann cell tubes, and find an end organ, the proximal nerve segment will form a neuroma. A neuroma consists of a disorganized, chaotic milieu of axons, Schwann cells, endoneurial cells, and perineurial cells within a dense myofibroblast stroma; up to 80% of the cross-sectional area of the neuroma consists of connective tissue (Cravioto and Battista 1981). Painful sensations can result within the distribution of the injured nerve. The literature suggests that between 3% and 30% of patients with neuromas experience symptomatic pain (Nelson 1977; Nashold et al. 1982; Atherton et al. 2006).

4.1.2 Mechanisms of Neuromatous Pain

Neuroma pain is caused by two general mechanisms: (1) persistent mechanical or chemical irritation of axons within the neuroma, and (2) the development of spontaneous, ectopic discharges within the neuroma and proximal retrograde axon sprouts. Normal pain sensations are caused by stimulation of faster conducting, lightly myelinated A-delta fibers and slower conducting, unmyelinated C fibers. In the normal physiologic state, these nociceptors generate action potentials only when subjected to high levels of stimuli. However, in the neuromatous state, inflammatory cells are released into the area, altering normal nociceptive pathways and creating abnormal neuropathic pain states. Pain may also result from abnormal connections between nociceptive fibers and the A-beta fibers that normally innervate skin. This contributes to a hyperexcitable, hyperalgesic state and results in spontaneous discharges within the abnormal nerve fibers. Hyperexcitability and ectopic discharges may also be caused by upregulation of sodium and potassium ion channels and adrenergic and nicotinic cholinergic receptors in axons within the neuroma (England et al. 1996; England et al. 1998; Ali et al. 2000; Pertin et al. 2005). Upregulation of these molecules creates a state of chemical sensitization that incites depolarizing currents and repeatedly generates action potentials, thereby creating neuropathic pain in the absence of external stimuli.

In a rat sciatic nerve model, Wall and Gutnick (1974) demonstrated ongoing spontaneous activity in smaller fibers within neuromas that may be responsible for pain sensations. In a study assessing neuromas of the superficial radial nerve in baboons, Meyer et al. (1985) also found that spontaneously active fibers were present, with apparent cross talk between fibers within the neuroma. These fibers consisted of both myelinated and unmyelinated axons that were mechanically sensitive. Sixty-seven percent of the spontaneously active fibers in the neuroma were unmyelinated, compared to 19% in the control. Since a large percentage of unmyelinated fibers in primates are nociceptive, this indicates a potential link between neuroma pain and spontaneous activity in noci-ceptive pathways.

4.1.3 Retrograde Changes in the Proximal Nerve Stump

Although neuromas themselves directly elicit neuropathic pain, axonal changes proximal to the neuroma may also contribute to pain sensations. Using a rat sciatic neuroma model, Amir and Devor (1993) showed that spontaneous discharges occurred both in afferents that terminated in the neuroma and in afferents with retrograde sprouts. In fact, spontaneous dis-charges were observed in 39% of fibers with retrograde sprouting, and axons with spontaneous activity were significantly more likely to have a retrograde sprout. The authors proposed that neurons with retrograde sprouts have an unusually high likelihood of firing spontaneously (Amir and Devor 1993), which, in conjunction with the increased capacity for myelinated A-beta sprouts to form connections with nociceptive fibers (Woolf et al. 1992; Woolf et al. 1995; Mannion et al. 1996), can result in pain.

4.1.4 Histologic Studies of Neuromas

To gain a better understanding of the proximal changes that occur in ampu-tated nerves, we performed a histomorphometric quantification of proximal axonal changes in a novel rabbit forelimb amputation model (Ko et al. 2011; Kim et al. 2012). Cross-sectional nerve specimens at various positions distal to the brachial plexus were harvested 8 weeks after forelimb amputation and nerve transection. Control cross sections from intact, contralateral nerves were harvested at equivalent positions. No significant differences were observed at different positions along the control nerves, so control nerve data for each nerve type were pooled. Histologic analysis demonstrated small, disorganized myelinated fibers with increased axonal sprouting toward the distal ends of the proximal transected nerve stumps. However, the myelin-ated fibers became progressively more organized at more proximal locations along the nerve stump (Figure 4.2). At the distal end, all transected nerves demonstrated statistically significant increases in nerve cross-sectional area compared to control nerve sections harvested from the intact contralateral side (Figure 4.3a).

The number of myelinated fibers progressively decreased in more proxi-mal sections, reaching normal levels 15 mm proximally, which corresponded to the level of the brachial plexus in our model (Figure 4.3b). The average cross-sectional area of a myelinated fiber was significantly decreased com-pared to controls in all sections of amputated nerves, indicating that atro-phic axonal changes proceeded proximally at least to the level of the brachial plexus (Figure 4.3c). In other words, we determined that after nerve transec-tion, morphologic changes at the axonal level extend beyond the region of gross neuroma formation in a distal-to-proximal fashion.

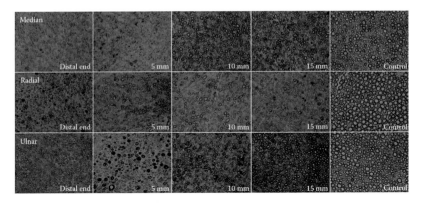

FIGURE 4.2
Toluidine blue stained sections of median nerve (*top row*), radial nerve (*center row*), and ulnar nerve (*bottom row*) at indicated distances proximal to the neuroma (400X magnification). *First column*: Smaller, disorganized myelinated fibers, with qualitatively increased amounts of myelin infolding, crenation, and debris are seen at the distal end of each proximal nerve stump. Regenerative clusters with axon sprouting are more prevalent at the distal ends, as is the amount of connective tissue stroma. *Second, third, and fourth columns*: The myelinated fibers become progressively more organized and larger at 5, 10, and 15 mm proximally, although myelin debris and crenation are still noted. *Fifth column*: Control nerve fibers are organized, circular, and larger with no noticeable myelin debris. (Adapted from Ko et al., A quantitative evaluation of gross versus histologic neuroma formation in a rabbit forelimb amputation model: Potential implications for the operative treatment and study of neuromas. *J Brachial Plex Peripher Nerve Inj* 6:8, 2011. With permission.)

4.2 Treatment of Neuromas

4.2.1 Nonsurgical Methods

Neuromas are usually most problematic when the patient wears a prosthesis and thus puts pressure on the neuroma, causing pain. Therefore, treatment usually starts with prosthetic modifications, including attempts to relieve pressure in tender areas by making adjustments to the socket. Increasing the loading forces on more tolerant areas, such as the tibial flairs in a transtibial socket, can also be helpful. The alignment of the prosthesis should be reassessed to ensure that there is not undue loading on the tender area; loading should be shifted away from this area if possible. Such adjustments are often adequate to alleviate the worst of the symptoms and enable the prosthesis to be worn comfortably.

Many other nonsurgical treatments, similar to those used for general neuropathic pain, can be tried, though they are usually of limited success. Nonpharmacologic treatments can include scar massage, desensitization techniques, biofeedback, and transcutaneous nerve stimulation. Pharmacologic treatments for neuroma pain include analgesics, such as nonsteroidal anti-inflammatory drugs (NSAIDs), acetaminophen, or narcotics—although we

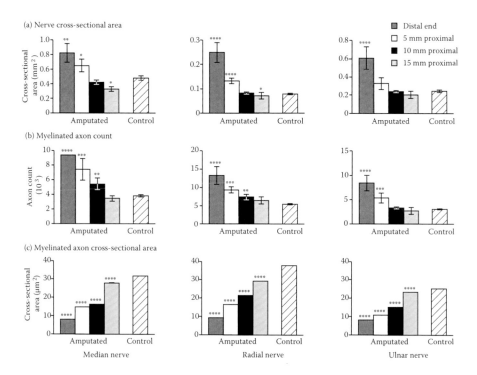

FIGURE 4.3

Differences in (a) nerve cross-sectional area; (b) myelinated axon count; and (c) myelinated axon cross-sectional area of the median, radial, and ulnar nerves at indicated distances from the distal nerve ending (6–8 weeks after amputation) compared to pooled non-amputated controls. Significant differences from controls are indicated: $*p = 0.05$; $**p = 0.01$; $***p = 0.001$; $****p = 0.0001$. (Adapted from Ko et al., A quantitative evaluation of gross versus histologic neuroma formation in a rabbit forelimb amputation model, *J Brachial Plex Peripher Nerve Inj* 6:8, 2011. With permission.)

recommend against using narcotics for long-term treatment of neuroma pain. Analgesic creams and topical patches are also used, with limited success. Anesthetics, such as lidocaine or bupivacaine, can be injected into the neuroma with or without steroids (Chabal et al. 1992). This can be useful in the diagnosis of a focal neuroma, though the pain relief is generally temporary. The addition of a small amount of steroid reduces local inflammation; however, we also find that this provides only a temporary benefit, if any. Neuropathic pain medications, thought to work primarily on central pain mechanisms (McQuay et al. 1996; Backonja and Glanzman 2003; Freynhagen et al. 2005), are also used to treat neuroma pain, although we have not found them to be effective. (For a recent review of neuropathic medications and their presumed mechanisms, see Gilron et al. 2006.) Neuroma pain is generated from a discrete, distal source, and this should be the focus of treatment.

Chemical axonotmesis, the injection of a sclerosing agent (e.g., phenol, glycerol, or alcohol) to cause focal axon degeneration, has demonstrated mixed results. Westerlund and colleagues demonstrated that intra- or extrafascicu-lar injection with phenol produced focal swelling of the nerve, severe demye-lination, axonal degeneration, edema, and hemorrhage, ultimately leading to complete disruption of nerve architecture (Westerlund et al. 2001). In a study using ultrasound guidance to direct phenol injections into neuromas in ten residual limbs, Gruber et al. (2004) reported successful resolution of pain in the early postinjection period. Radio-frequency ablation can also be used to destroy or damage the nerve fibers in a neuroma and is used as a treatment for neuroma pain (Tamimi et al. 2009; Restrepo-Garces et al. 2011). In our experience, both sclerosing agents and ablation techniques have produced inconsistent results; using either of these approaches results in high rates of neuroma recurrence. We hypothesize that although sclerosing agents and radiofrequency ablation destroy the neuroma, any relief is temporary, as the injured proximal axons then regrow to form another neuroma. Therefore, we no longer recommend these treatment modalities.

4.2.2 Surgical Methods

If nonsurgical methods are unsuccessful and neuroma pain persists, surgi-cal intervention is generally indicated. However, there are at least 150 surgi-cal treatments for neuromas that have been described in the literature (Wood and Mudge 1987); this myriad of proposed treatments highlights the fact that no single surgical option has proven superior.

Simply excising the neuroma without any further manipulation yields a high likelihood of recurrence of the neuroma. The normal physiologic response of severed nerve fibers is to regenerate and reinnervate a target organ; if functional end organs are not available, the neuroma will always reform. Placing gentle traction on the nerve while excising the neuroma (traction neurectomy) allows the nerve to retract more proximally. The inten-tion is to place the inevitable end-neuroma away from superficial locations in which it is more likely to become irritated.

Numerous studies have described covering the resected proximal nerve stump with caps of various materials (including silicone) to contain the neuroma and protect it from inflammatory changes in the surrounding tis-sues (Tupper and Booth 1976; Swanson et al. 1977; Muehleman and Rahimi 1990; Sakai et al. 2005), but the results have been consistently poor. Other techniques involve end-to-side or centro-central coaptation of the affected nerve (González-Darder et al. 1985; Ashley and Stallings 1988; Barberá and Albert-Pampló 1993), while others advocate covering the neuromas with vas-cularized tissue flaps (Krishnan et al. 2005). The translocation of nerves into a hole drilled into bone as a way to limit axon growth is not a new con-cept and has demonstrated promise, especially in the hand (Boldrey 1943; Mass et al. 1984; Hazari and Elliot 2004). However, our experience with this technique is that patients tend to do well for approximately 1 year, but then their pain recurs. We hypothesize that new bone grows into the bone-nerve interface over time, ultimately compressing the implanted nerve and causing pain; therefore, we no longer recommend this approach. Other authors have described transposing a transected nerve ending into the end of a cut vein (Herbert and Filan 1998; Mobbs et al. 2003; Balcin et al. 2009), which provides the advantage that subcutaneous veins are readily available throughout the body. By placing a vein over a cut nerve ending, the vein essentially acts as a nerve cap that protects the nerve from its surrounding environment while perhaps actively transporting neurotrophic factors away from the nerve ending. In one study, Balcin and colleagues demonstrated that this technique resulted in less pain than transposing the nerve into muscle (Balcin et al. 2009). However, excision of the neuroma and transposition of the nerve ending into muscle—that is, burying the nerve in muscle—remains the "gold standard" for neuroma treatment.

Dellon and colleagues demonstrated that after excision of neuromas and transposition of the nerve into muscle, nerve endings had a different histologic appearance than non-transposed nerve endings. There was no interaction with muscle fibers, but the random, disorganized whorls of neural fibers and dense connective tissue that characterize the classic neuroma were absent (Dellon et al. 1984). Mackinnon et al. (1985) reported that biopsies of nerves implanted into muscle demonstrated no evidence of nerve regeneration through muscle into overlying tissue. Dellon and Mackinnon followed up these observations with a case series in which 78 neuromas were excised and the nerves buried in muscle. In this study, 81% of the patients had either good (i.e., improved pain; 39%) or excellent (no pain; 42%) results (Dellon and Mackinnon 1986). It was assumed in these studies that after burying the cut nerve ending in muscle, any recurrent neuroma is cushioned by the muscle and thus is less likely to be subject to repeat trauma or mechanical stimulation, thus reducing pain.

4.3 TMR as a Strategy for Neuroma Prevention

4.3.1 Theory

The central principle of TMR surgery involves hyper-reinnervation of denervated target muscles to create new "myoneurosomes" (Hijjawi et al. 2006) that allow intuitive, seamless control of a prosthesis (Kuiken et al. 1995; Kuiken et al. 2004). Early in a clinical series of TMR procedures, we observed that although several of the amputees complained of neuroma pain prior to TMR surgery, the majority of these patients reported resolution of their neuroma pain symptoms after TMR. Our hypothesis was that by directing the amputated nerve stumps onto motor points of *denervated* muscle, TMR allows regenerating axons to grow into the target muscle instead of forming a chaotic, disorganized neuroma.

4.3.2 Basic Scientific Evidence

In an early study, we documented neuroma formation in the terminal branches of the rabbit brachial plexus after forelimb amputation (Kim et al. 2010). Using this animal model, we transferred previously amputated nerve stumps, after excision of end-neuromas, to a denervated, pedicled rectus abdominis muscle flap. Ten weeks after transfer, the nerve morphology more closely resembled that of the non-amputated controls than that of a neuroma (Figure 4.4) (Kim et al. 2012). Histomorphometric analysis demonstrated a significant decrease in the number of myelinated fibers (Figure 4.5a), along with a similar increase in the average myelinated fiber diameter (Figure 4.5b) compared to non-transferred nerve endings (neuroma). In other words, after TMR, the histologic parameters of the transferred nerves more closely resembled those of uninjured controls. Therefore, TMR appeared to effectively prevent the formation of new neuromas.

To further test the effects of TMR on neuroma formation, we employed a rat hind limb model, in which we evaluated the effects of TMR transfers in mixed (common peroneal) and pure sensory (sural) nerves. In the common peroneal *treatment group,* the common peroneal nerve was transferred to the motor point of the denervated medial gastrocnemius muscle. In the common peroneal *control group A,* the common peroneal nerve was transposed into the innervated medial gastrocnemius muscle away from the motor end point (i.e., buried in muscle). In the common peroneal *control group B,* the common peroneal nerve was transferred to the motor end point of an innervated medial gastrocnemius muscle (Figure 4.6a). Similar procedures were also performed for the purely sensory sural nerve: the sural nerve was transferred either to the motor point of denervated medial gastrocnemius (*sural treatment group*) or buried in innervated muscle away from the motor point (*sural control group*) (Figure 4.6b).

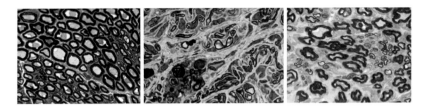

FIGURE 4.4
Longitudinal section of a single median nerve before transection (*left*), 6 weeks after neuroma formation (*center*), and 10 weeks post-TMR (*right*). These histologic images (obtained at 100X) qualitatively demonstrate the effect of TMR on neuromas—namely, a decrease in myelinated fiber density and an increase in the average myelinated fiber size. TMR reduces the histologic changes seen in neuroma formation. (Reprinted and adapted from Kim et al., The effects of targeted muscle reinnervation on neuromas in a rabbit rectus abdominis flap muscle, *J Hand Surg [Am]* 37 (8):1609–1616. © [2012]. With permission from Elsevier.)

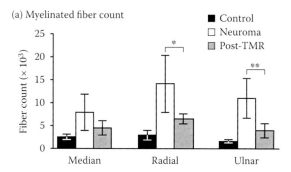

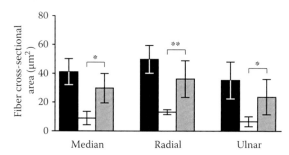

FIGURE 4.5

(a) Myelinated fiber count and (b) cross-sectional areas of myelinated fibers in the median (n = 5 control, 5 neuroma, 4 post-TMR), radial (n = 5 control, 4 neuroma, 3 post-TMR), and ulnar (n = 4 control, 4 neuroma, 3 post-TMR) nerves. Post-TMR samples significantly different from neuroma samples are indicated: *p = 0.05; **p = 0.02. (Reprinted and adapted from Kim et al., The effects of targeted muscle reinnervation on neuromas in a rabbit rectus abdominis flap muscle, *J Hand Surg [Am]* 37 (8):1609–1616. © [2012]. With permission from Elsevier.)

Sixteen weeks after the TMR procedures, the animals in the common peroneal treatment group demonstrated significantly decreased gross neuroma formation compared to common peroneal control groups A and B. Likewise, the sural nerve treatment group exhibited significantly decreased gross neuroma formation compared to controls (Figure 4.7). Nerves from the common peroneal and sural treatment groups also displayed significantly decreased histologic evidence of neuroma formation compared to their respective control groups (Figures 4.8, 4.9, 4.10). Our study indicates that transferring either mixed or purely sensory nerves into denervated muscle motor points (i.e., performing TMR) results in decreased neuroma formation compared to simply transposing the nerves into muscle, which is considered by many to be the current gold standard treatment for neuromas.

These data provided us with an intuitive hypothesis—namely, that impairing the normal regenerative pathways results in neuroma formation and that this can be prevented by providing the severed nerve with a target. Transferring an amputated nerve to a *denervated* motor end point

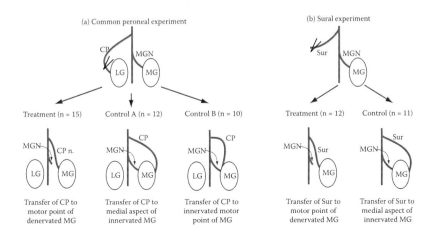

FIGURE 4.6
(a) The mixed sensory/motor common peroneal (CP) nerve was transferred either to the denervated medial gastrocnemius (MG) muscle in the common peroneal treatment group ($n = 15$); to the innervated MG muscle at a site distant from the motor end point in the common peroneal control group A ($n = 12$); or to the motor end point of an innervated MG muscle in the common peroneal control group B ($n = 10$). (b) In the sural treatment group ($n = 12$), the purely sensory sural nerve (Sur) was transferred to the denervated MG muscle (sural treatment group); in the sural control group, the sural nerve was transferred away from the motor end point on the innervated MG muscle ($n = 11$).

provides the nerve with a small but important distal nerve segment and the necessary end organs. The denervated distal nerve segment and muscle presumably provide the necessary neurotrophic factors and act as conduits for axon regeneration. We hypothesize that when distal nerve tubes, neurotrophic factors, and distal targets are provided for the regenerating nerve, regeneration will progress to a quiescent state, rather than to a disorganized, chaotic neuroma.

This hypothesis is further supported by earlier studies in which denervated rat gastrocnemius muscle was hyper-reinnervated with one of several combinations of hind limb nerves—an experimental model very similar to that described above (Kuiken et al. 1995; see Chapter 2). Robust reinnervation occurred: transferred nerves formed up to three times as many motor units as found in the natively innervated muscle, and muscle mass and function recovered to normal levels. Thus, after transfer to a motor point on denervated muscle, motor axons actively regenerated into the muscle and formed functional connections with the medial gastrocnemius muscle fibers.

Interestingly, in the above experiments, neuroma formation in the purely sensory sural nerve also appears to be mitigated after transfer to muscle. Presumably, trophic factors and scaffolding provided by the distal nerve segment are sufficient to allow sensory nerve growth. In TMR, brachial plexus nerves that contain both motor and sensory fibers are transferred to muscle.

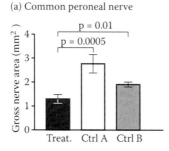

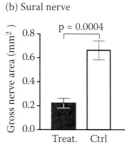

(a) Common peroneal nerve   (b) Sural nerve

FIGURE 4.7
(a) Sixteen weeks after the TMR procedures, nerves in the common peroneal treatment group (Treat.) demonstrated significantly decreased gross neuroma formation compared to common peroneal control group A (Ctrl A) and common peroneal control group B (Ctrl B), as determined by micro-caliper measurements of gross nerve area. (b) Similarly, the sural treatment group exhibited significantly decreased gross neuroma formation compared to sural controls ($p = 0.0004$).

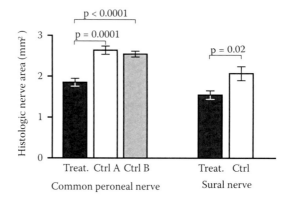

FIGURE 4.8
Histologically, the common peroneal treatment animals (Treat.) displayed significantly decreased neuroma formation compared to the common peroneal control group A (Ctrl A) and common peroneal control group B (Ctrl B), as determined by quantitative longitudinal measurements (area of the distal 5 mm of neural tissue at the nerve-muscle interface). The sural treatment animals (Treat.) also demonstrated decreased neuroma formation histologically when compared to sural control animals.

In several TMR patients, sensory nerves have regenerated through target muscle to reach cutaneous effectors, providing patients with the perception of sensory feedback from their missing limb, a phenomenon termed transfer sensation (Kuiken et al. 2007a; Kuiken et al. 2007b; Schultz et al. 2009; Marasco et al. 2009; Marasco et al. 2011). (See Chapter 8 for a detailed description of sensory reinnervation.)

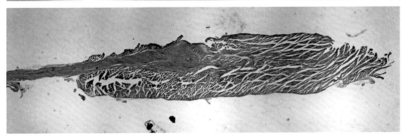

FIGURE 4.9
In longitudinal sections stained with hematoxylin and eosin (H&E), common peroneal treatment nerves (*top*) demonstrated decreased longitudinal diameters compared to common peroneal control group A (*center*) and common peroneal control group B (*bottom*) nerves. In addition to the increased sizes of the control nerves, note the whorls of axons and perineurial cells intertwined with tangles of connective tissue stroma, consistent with neuroma histology, which are seen in the common peroneal control group A and common peroneal control group B nerves, but not in the common peroneal treatment group (20X).

4.3.3 Clinical Evidence

At Northwestern Memorial Hospital (NMH),* in conjunction with the Center for Bionic Medicine (CBM) at the Rehabilitation Institute of Chicago (RIC), our clinical experience with TMR has been consistent with the favorable effects of TMR on neuroma formation seen in our animal studies. As evident in Table 4.1, five out of nine shoulder disarticulation patients had neuroma pain prior to their TMR procedure; however, no patients complained of

---

* All clinical data presented were obtained with consent of the Institutional Review Boards from the appropriate institutions.

FIGURE 4.10
In longitudinal sections stained with H&E, sural treatment nerves (*top*) demonstrated decreased longitudinal diameters and decreased amounts of neuromatous whorling and interlacing bundles compared to sural control nerves (*bottom*) (20X).

neuroma pain after TMR surgery. In other words, by transferring amputated nerve endings into denervated target muscles, we were able to eliminate all preexisting neuroma pain in these patients.

Of eight transhumeral patients, four had preexisting neuroma pain; after TMR surgery, three of those four continued to complain of neuroma pain. However, further analysis indicated that two of the patients had pain due to neuromas in the musculocutaneous nerve (Patient 14) or lateral antebrachial cutaneous nerve (Patient 15). Neither of these nerves had been transferred to denervated muscle. Patient 16 continued to experience post-TMR neuroma pain, although the postoperative pain level was significantly reduced compared to preoperative levels.

Harborview Medical Center at the University of Washington in Seattle* is the only Level I trauma center for five surrounding states. Consequently, a high volume of patients with traumatic amputations (and injuries requiring revision amputations of an affected limb) are transferred to our center for acute care. Over several years, our group has been performing TMR specifically to prevent or treat neuroma pain, especially in the lower limb amputee population. We have performed TMR both in the acute setting (i.e., once the soft tissues had stabilized, usually within days of the amputation) and the nonacute setting (i.e., several months to years after amputation). Although there is undoubtedly concern that the zone of injury to the nerve may be

---

* All clinical data presented were obtained with consent of the Institutional Review Boards from the appropriate institutions.

TABLE 4.1

Effects of Targeted Muscle Reinnervation (TMR) on Neuroma Pain. (TMR performed at Northwestern Memorial Hospital [NMH], Chicago; Rehabilitation performed at the Rehabilitation Institute of Chicago [RIC].)

Patient No.	Amputation Level	Duration between Amputation and TMR (Mo.)	Pre-TMR Neuroma Pain	Duration of Follow-up (Mo.)	Post-TMR Neuroma Pain
1	Shoulder disarticulation	10	Yes	124	No
2	Shoulder disarticulation	14	Yes	82	No
3	Shoulder disarticulation	15	No	6	No
4	Shoulder disarticulation	9	No	11	No
5	Shoulder disarticulation	5	Yes	7	No
6	Shoulder disarticulation	21	No	23	No
7	Shoulder disarticulation	17	Yes	10	No
8	Shoulder disarticulation	8	Yes	8	No
9	Shoulder disarticulation	16	No	4	No
10	Transhumeral, long	12	No	6	No
11	Transhumeral, long	13	Yes	30	No
12	Transhumeral, long	9	No	N/A	No
13	Transhumeral, mid	6	No	60	No
14[a]	Transhumeral, long	12	Yes	23	Yes
15[b]	Transhumeral, long	73	Yes	18	Yes
16[c]	Transhumeral, short	18	Yes	15	Yes
17	Transhumeral, long	7	No	5	No

[a] Musculocutaneous neuroma, nerve not transferred.
[b] Lateral antebrachial cutaneous neuroma, nerve not transferred.
[c] Neuroma pain improved post-TMR.

underestimated when performing nerve transfers in the acute setting (Ko et al. 2011), we believe that excising the nerve stumps as proximally as possible will minimize formation of neuromas at the nerve-muscle interface. As shown in Table 4.2, we have follow-up data for a series of 32 TMR patients: 14 upper extremity TMR patients (9 acute and 5 nonacute) and 18 lower extremity patients (2 acute, 16 nonacute).

Of the 21 patients who had nonacute TMR surgery, all complained of pre-TMR neuroma pain; however, none experienced post-TMR neuroma pain, indicating a marked reduction in symptomatic neuroma formation after TMR surgery.

We have performed 11 TMR procedures in the acute traumatic amputation setting. Of these patients, only two developed mild neuroma pain. Our experience leads us to believe that performing TMR in the acute amputation setting may be an effective method to prevent postamputation neuroma pain; however, further study is necessary and is ongoing.

TABLE 4.2

Effects of Targeted Muscle Reinnervation (TMR) on Neuroma Pain. (TMR performed at Harborview Medical Center, University of Washington, Seattle, Washington.)

Patient	Amputation Level	Nerve Transfers[a]	Duration (mo)		Neuroma	
			Injury to TMR[b]	Follow-up	Pre-TMR	Post-TMR
1	Shoulder disarticulation		10	5	Yes	No
2[c]	Shoulder disarticulation	Median to pectoralis, radial to teres major, ulnar to serratus	0	11	N/A	Yes
3	Transhumeral, long		0	8	N/A	No
4	Transhumeral, long		0	5	N/A	No
5	Transhumeral, long		0	9	N/A	No
6	Transhumeral, long		0	14	N/A	No
7	Transhumeral, long		77	27	Yes	No
8	Transhumeral, mid		52	6	Yes	No
9	Transhumeral, short		0	57	N/A	No
10	Transhumeral, short	Musculocutaneous to medial biceps, median to lateral biceps, radial to distal triceps, ulnar to latissimus dorsi	0	16	N/A	No
11	Transhumeral, short	Median to upper pectoralis, radial to serratus, ulnar to lower pectoralis	0	48	N/A	Yes
12	Transhumeral, short	Median to pectoralis, radial to latissimus dorsi, ulnar to teres major	0	23	N/A	No
13	Transradial	Radial to supinator and abductor pollicis longus	144	53	Yes	No
14	Transradial	Median to brachialis, ulnar to triceps	30	9	Yes	No
15	Above-knee	Tibial to medial hamstring, peroneal to lateral hamstring	0	32	No	No

TABLE 4.2 (continued)

Effects of Targeted Muscle Reinnervation (TMR) on Neuroma Pain. (TMR performed at Harborview Medical Center, University of Washington, Seattle, Washington.)

Patient	Amputation Level	Nerve Transfers[a]	Duration (mo)		Neuroma	
			Injury to TMR[b]	Follow-up	Pre-TMR	Post-TMR
16	Above-knee	Tibial to medial hamstring, peroneal to lateral hamstring	138	48	Yes	No
17	Above-knee	Sciatic to lateral hamstring	351	4	Yes	No
18	Above-knee	Sciatic to hamstring	15	33	Yes	No
19	Above-knee	Sciatic to medial hamstring	25	4	Yes	No
20[d]	Above-knee	Sciatic to hamstring	48	3	Yes	No
21	Knee disarticulation	Tibial to medial hamstring, peroneal to lateral hamstring	0	12	N/A	No
22	Below-knee	Tibial to biceps femoris	53	3	Yes	No
23	Below-knee	Sciatic to lateral hamstring	57	23	Yes	No
24	Below-knee	Deep and superficial peroneal to peroneus brevis	46	48	Yes	No
25	Below-knee	Deep and superficial peroneal to tibialis anterior	128	19	Yes	No
26	Below-knee	Tibial to medial hamstring, peroneal to lateral hamstring	N/A[e]	3	Yes	No
27	Below-knee	Tibial to medial hamstring, peroneal to lateral hamstring	292	12	Yes	No
28	Below-knee	Tibial to medial hamstring, peroneal to lateral hamstring	53	41	Yes	No
29	Below-knee	Peroneal to lateral hamstring	269	13	Yes	No
30	Below-knee	Tibial to medial hamstring, peroneal to lateral hamstring	N/A	4	Yes	No
31	Below-knee	Tibial to medial hamstring	83	5	Yes	No

(continued)

TABLE 4.2 (continued)

Effects of Targeted Muscle Reinnervation (TMR) on Neuroma Pain. (TMR performed at Harborview Medical Center, University of Washington, Seattle, Washington.)

Patient	Amputation Level	Nerve Transfers[a]	Duration (mo)		Neuroma	
			Injury to TMR[b]	Follow-up	Pre-TMR	Post-TMR
32	None	Sural to flexor hallucis longus	127	7	Yes	No

[a] Nerve transfers only specified if atypical coaptations performed.
[b] Duration of 0 months indicates acute TMR.
[c] Patient experienced mild post-TMR neuroma pain.
[d] Patient's severe complex regional pain syndrome was successfully treated with above-knee amputation and TMR.
[e] Patient had congenital neuromas.

Our results suggest that nerve transfers to denervated muscle targets can be considered solely to help prevent neuroma formation, even if the transfer is not intended to create new control signals. It should be noted, however, that for functional TMR, the surgical procedures outlined in Chapter 3 should be followed to optimize the myoelectric signals for prosthesis control.

4.4 Conclusion

By directing amputated nerve stumps into denervated muscle, TMR allows regenerating axons to grow into the target muscle instead of forming a chaotic, disorganized neuroma. Our animal experiments and translational experience indicate that TMR is an effective method of treating preexisting painful neuromas in amputees. In addition, TMR has demonstrated significant potential as a means to prevent neuroma formation in the acute injury setting, although further research is warranted. TMR may represent a paradigm shift in how we both prevent and treat neuromas in amputees and in other patient populations.

References

Ali, Z., S. N. Raja, U. Wesselmann, P. N. Fuchs, R. A. Meyer, and J. N. Campbell. 2000. Intradermal injection of norepinephrine evokes pain in patients with sympathetically maintained pain. *Pain* 88:161–168.

Amir, R., and M. Devor. 1993. Ongoing activity in neuroma afferents bearing retrograde sprouts. *Brain Res* 1630:283–288.

Ashley, L., and J. O. Stallings. 1988. End-to-side nerve flap for treatment of painful neuroma: a 15-year follow-up. *J Am Osteopath Assoc* 88:621–624.

Atherton, D. D., O. Taherzadeh, P. Facer, D. Elliot, and P. Anand. 2006. The potential role of nerve growth factor (NGF) in painful neuromas and the mechanism of pain relief by their relocation to muscle. *J Hand Surg [Br]* 31:652–656.

Backonja, M., and R. L. Glanzman. 2003. Gabapentin dosing for neuropathic pain: evidence from randomized, placebo-controlled clinical trials. *Clin Ther* 25:81–104.

Balcin, H., P. Erba, R. Wettstein, D. J. Schaefer, G. Pierer, and D. F. Kalbermatten. 2009. A comparative study of two methods of surgical treatment for painful neuroma. *J Bone Joint Surg Br* 91:803–808.

Barberá, J., and R. Albert-Pampló. 1993. Centrocentral anastomosis of the proximal nerve stump in the treatment of painful amputation neuromas of major nerves. *J Neurosurg* 79:331–334.

Boldrey, E. 1943. Amputation neuroma in nerves implanted in bone. *Ann Surg* 118:1052–1057.

Chabal, C., L. Jacobson, L. C. Russell, and K. J. Burchiel. 1992. Pain response to perineuromal injection of normal saline, epinephrine, and lidocaine in humans. *Pain* 49:9–12.

Cravioto, H., and A. Battista. 1981. Clinical and ultrastructural study of painful neuroma. *Neurosurgery* 8:181–190.

Dellon, A. L., S. E. Mackinnon, and A. Pestronk. 1984. Implantation of sensory nerve into muscle: preliminary clinical and experimental observations on neuroma formation. *Ann Plast Surg* 12:30–40.

Dellon, A. L., and S. E. Mackinnon. 1986. Treatment of the painful neuroma by neuroma resection and muscle implantation. *Plast Reconstr Surg* 77:427–438.

England, J. D., L. T. Happel, D. G. Kline, et al. 1996. Sodium channel accumulation in humans with painful neuromas. *Neurology* 47:272–276.

England, J. D., L. T. Happel, Z. P. Liu, C. L. Thouron, and D. G. Kline. 1998. Abnormal distributions of potassium channels in human neuromas. *Neurosci Lett* 255:37–40.

Freynhagen, R., K. Strojek, T. Griesing, E. Whalen, and M. Balkenohl. 2005. Efficacy of pregabalin in neuropathic pain evaluated in a 12-week, randomized, double-blind, multicenter, placebo-controlled trial of flexible- and fixed-dose regimens. *Pain* 115:254–263.

Gilron, I., C. P. Watson, C. M. Cahill, and D. E. Moulin. 2006. Neuropathic pain: a practical guide for the clinician. *CMAJ* 175:265–275.

González-Darder, J., J. Barberá, M. J. Abellán, and A. Mora. 1985. Centrocentral anastomosis in the prevention and treatment of painful terminal neuroma. An experimental study in the rat. *J Neurosurg* 63:754–758.

Gruber, H., P. Kovacs, S. Peer, B. Frischhut, and G. Bodner. 2004. Sonographically guided phenol injection in painful stump neuroma. *AJR Am J Roentgenol* 182:952–954.

Hazari, A., and D. Elliot. 2004. Treatment of end-neuromas, neuromas-in-continuity and scarred nerves of the digits by proximal relocation. *J Hand Surg [Br]* 29:338–350.

Herbert, T. J., and S. L. Filan. 1998. Vein implantation for treatment of painful cutaneous neuromas. A preliminary report. *J Hand Surg Br* 23:220–224.

Hijjawi, J. B., T. A. Kuiken, R. D. Lipschutz, L. A. Miller, K. A., Stubblefield, and G. A. Dumanian. 2006. Improved myoelectric prosthesis control accomplished using multiple nerve transfers. *Plast Reconstr Surg* 118 (7): 1573–1578.

Kim, P. S., J. Ko, K. O'Shaughnessy, T. A. Kuiken, and G. A. Dumanian. 2010. Novel model for end-neuroma formation in the amputated rabbit forelimb. *J Brachial Plex Peripher Nerve Inj* 5:6.

Kim, P. S., J. H. Ko, K. D. O'Shaughnessy, T. A. Kuiken, E. A. Pohlmeyer, and G. A. Dumanian. 2012. The effects of targeted muscle reinnervation on neuromas in a rabbit rectus abdominis flap model. *J Hand Surg [Am]* 37 (8):1609–1616.

Ko, J. H., P. S. Kim, K. D. O'Shaughnessy, X. Ding, T. A. Kuiken, and G. A. Dumanian. 2011. A quantitative evaluation of gross versus histologic neuroma formation in a rabbit forelimb amputation model: potential implications for the operative treatment and study of neuromas. *J Brachial Plex Peripher Nerve Inj* 6:8.

Krishnan, K. G., T. Pinzer, and G. Schackert. 2005. Coverage of painful peripheral nerve neuromas with vascularized soft tissue: method and results. *Neurosurgery* 56:369–378.

Kuiken, T. A., D. S. Childress, and W. Z. Rymer. 1995. The hyper-reinnervation of rat skeletal muscle. *Brain Res* 676:113–123.

Kuiken, T. A., G. A. Dumanian, R. D. Lipschutz, L. A. Miller, and K. A. Stubblefield. 2004. The use of targeted muscle reinnervation for improved myoelectric prosthesis control in a bilateral shoulder disarticulation amputee. *Prosthet Orthot Int* 28:245–253.

Kuiken, T. A., P. D. Marasco, B. A. Lock, R. N. Harden, and J. P. A. Dewald. 2007a. Redirection of cutaneous sensation from the hand to the chest skin of human amputees with targeted reinnervation. *Proc Natl Acad Sci USA* 104 (50):20061–20066.

Kuiken, T. A., L. A. Miller, R. D. Lipschutz, et al. 2007b. Targeted reinnervation for enhanced prosthetic arm function in a woman with a proximal amputation: a case study. *Lancet* 369 (9559):371–380.

Mackinnon, S. E., A. L. Dellon, A. R. Hudson, and D. A. Hunter. 1985. Alteration of neuroma formation by manipulation of its microenvironment. *Plast Reconstr Surg* 76:345–353.

Mannion, R. J., T. P. Doubell, R. E. Coggeshall, and C. J. Woolf. 1996. Collateral sprouting of uninjured primary afferent A-fibers into the superficial dorsal horn of the adult rat spinal cord after topical capsaicin treatment to the sciatic nerve. *J Neurosci* 16:5189–5195.

Marasco, P. D., K. Kim, J. E. Colgate, M. A. Peshkin, and T. A. Kuiken. 2011. Robotic touch shifts perception of embodiment to a prosthesis in targeted reinnervation amputees. *Brain* 134 (Pt 3):747–758.

Marasco, P. D., A. E. Schultz, and T. A. Kuiken. 2009. Sensory capacity of reinnervated skin after redirection of amputated upper limb nerves to the chest. *Brain* 132 (6):1441–1448.

Mass, D. P., M. C. Ciano, R. Tortosa, W. L. Newmeyer, and E. S. Kilgore Jr. 1984. Treatment of painful hand neuromas by their transfer into bone. *Plast Reconstr Surg* 74:182–185.

McQuay, H. J., M. Tramer, B. A. Nye, D. Carroll, P. J. Wiffen, and R. A. Moore. 1996. A systematic review of antidepressants in neuropathic pain. *Pain* 68:217–227.

Meyer, R. A., S. N. Raja, J. N. Campbell, S. E. Mackinnon, and A. L. Dellon. 1985. Neural activity originating from a neuroma in the baboon. *Brain Res* 325:255–260.

Mobbs, R. J., M. Vonau, and P. Blum. 2003. Treatment of painful peripheral neuroma by vein implantation. *J Clin Neurosci* 10:338–339.

Muehleman, C., and F. Rahimi. 1990. Effectiveness of an epineurial barrier in reducing axonal regeneration and neuroma formation in the rat. *J Foot Surg* 29:260–264.

Nashold, B. S., Jr., J. L. Goldner, J. B. Mullen, and D. S. Bright. 1982. Long-term pain control by direct peripheral-nerve stimulation. *J Bone Joint Surg Am* 64:1–10.

Nelson, A. W. 1977. The painful neuroma: the regenerating axon verus the epineural sheath. *J Surg Res* 23:215–221.

Odier, L. 1811. *Manuel de médecine pratique.* Geneva: J. J. Paschoud.

Paré, A. 1634. *The Collected Works of Ambroise Paré.* London: T. Cotes and R. Young.

Pertin, M., R. R. Ji, T. Berta, et al. 2005. Upregulation of the voltage-gated sodium channel beta2 subunit in neuropathic pain models: characterization of expression in injured and non-injured primary sensory neurons. *J Neurosci* 25:10970–10980.

Ramón y Cajal, S. 1905. Mecanismo de la degeneration y regeneration de nervos. *Trab Lab Inbest Biol* 9:119.

Restrepo-Garces, C. E., A. Marinov, P. McHardy, G. Faclier, and A. Avila. 2011. Pulsed radiofrequency under ultrasound guidance for persistent stump-neuroma pain. *Pain Pract* 11:98–102.

Sakai, Y., M. Ochi, Y. Uchio, K. Ryoke, and S. Yamamoto. 2005. Prevention and treatment of amputation neuroma by an atelocollagen tube in rat sciatic nerves. *J Biomed Mater Res B Appl Biomater* 73:355–360.

Schultz, A. E., P. D. Marasco, and T. A. Kuiken. 2009. Vibrotactile detection thresholds for chest skin of amputees following targeted reinnervation surgery. *Brain Res* 1251:121–129.

Swanson, A. B., N. R. Boeve, and R. M. Lumsden. 1977. The prevention and treatment of amputation neuromata by silicone capping. *J Hand Surg Am* 2:70–78.

Tamimi, M. A., M. H. McCeney, and J. Krutsch. 2009. A case series of pulsed radiofrequency treatment of myofascial trigger points and scar neuromas. *Pain Med* 10:1140–1143.

Tupper, J. W., and D. M. Booth. 1976. Treatment of painful neuromas of sensory nerves in the hand: a comparison of traditional and newer methods. *J Hand Surg [Am]* 1:144–151.

Virchow, R. 1863. *Die krankhaften Geschwülste.* Berlin: A. Hirschwald

Wall, P. D., and M. Gutnick. 1974. Ongoing activity in peripheral nerves: the physiology and pharmacology of impulses originating from a neuroma. *Exp Neurol* 43:580–593.

Waller, A. 1850. Experiments on the sections of glossopharyngeal and hypoglossal nerves of the frog and observations of the alterations produced thereby in the structure of their primitive fibers. *Philos Trans R Soc Lond* 140:423–429.

Westerlund, T., V. Vuorinen, and M. Roytta. 2001. Same axonal regeneration rate after different endoneurial response to intraneural glycerol and phenol injection. *Acta Neuropathol* 102:41–54.

Wood, V. E., and M. K. Mudge. 1987. Treatment of neuromas about a major amputation stump. *J Hand Surg [Am]* 12:302–306.

Wood, W. 1828. Observations on neuroma. *Trans Med-Chir Soc Edin* 29:367–433.

Woolf, C. J., P. Shortland, and R. E. Coggeshall. 1992. Peripheral nerve injury triggers central sprouting of myelinated afferents. *Nature* 355:75–78.

Woolf, C. J., P. Shortland, M. Reynolds, J. Ridings, T. Doubell, and R. E. Coggeshall. 1995. Reorganization of central terminals of myelinated primary afferents in the rat dorsal horn following peripheral axotomy. *J Comp Neurol* 360:121–134.

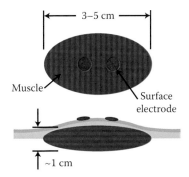

FIGURE 2.3
Minimum size of target muscle required for detection of adequate EMG signal strength after TMR.

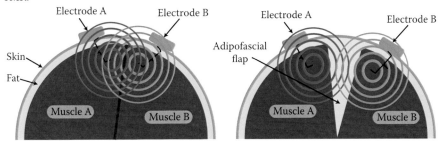

FIGURE 2.5
Use of an adipofascial flap to physically separate adjacent muscles or muscle segments. *Left*: Cross talk is caused by interference of EMG signal from adjacent muscles: electrode A (on muscle A) picks up EMG signals from muscle B, and electrode B picks up EMG signals from muscle A. *Right*: Physical separation of muscles A and B by an adipofascial flap reduces the cross talk detected at both electrodes A and B. EMG signals (*concentric blue circles*) are shown propagating from a point source, for clarity.

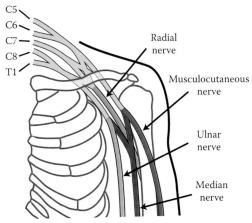

FIGURE 3.1
Diagram of brachial plexus anatomy.

(a) Key muscles for transhumeral TMR

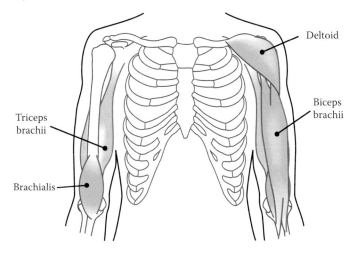

Deltoid

Biceps
brachii

Triceps
brachii

Brachialis

(b) TMR for transhumeral amputation

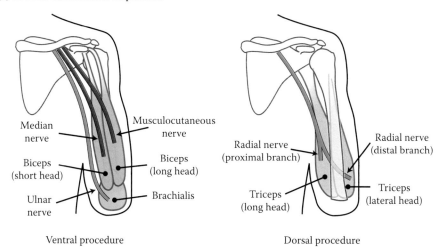

Median
nerve

Musculocutaneous
nerve

Biceps
(short head)

Biceps
(long head)

Ulnar
nerve

Brachialis

Radial nerve
(proximal branch)

Radial nerve
(distal branch)

Triceps
(long head)

Triceps
(lateral head)

Ventral procedure

Dorsal procedure

FIGURE 3.2
(a) Normal anatomy of key muscles for transhumeral TMR surgery. (b) Schematic of typical
surgical plan for transhumeral TMR.

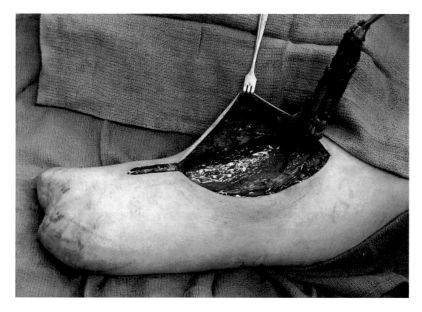

FIGURE 3.3

An adipofascial flap is raised during ventral dissection of the residual limb in a transhumeral amputee. Following nerve coaptation, the flap is placed between the long and short heads of the biceps brachii to facilitate isolation of the two EMG signals.

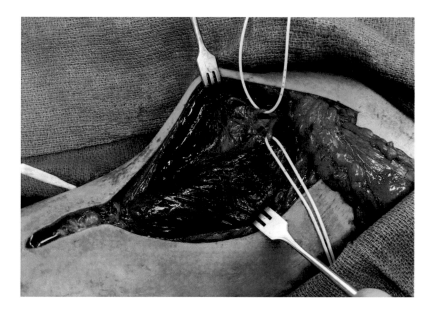

FIGURE 3.4

Ventral exposure revealing the motor branches of the musculocutaneous nerve to the long and short heads of the biceps.

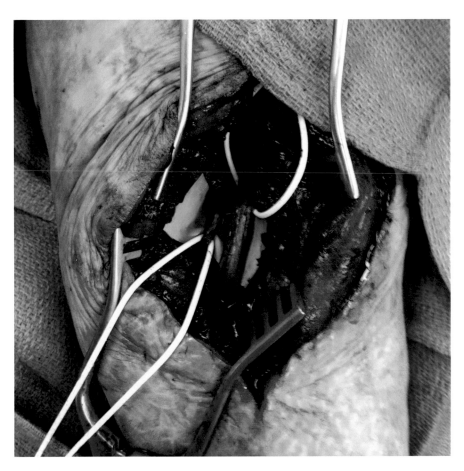

FIGURE 3.6
Dorsal exposure revealing the radial nerve and its motor branch to the lateral head of the triceps muscle.

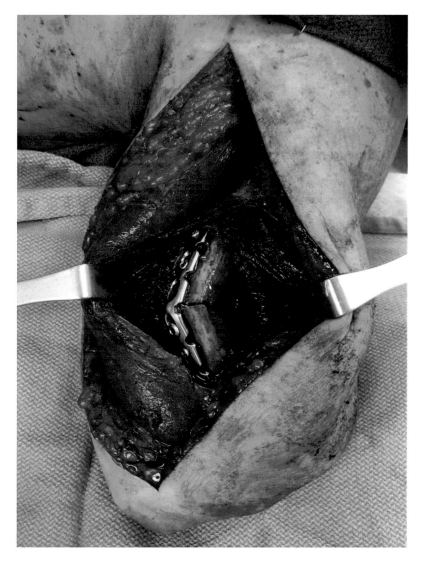

FIGURE 3.7
(a) Anterior view of angulation osteotomy in a transhumeral amputee.

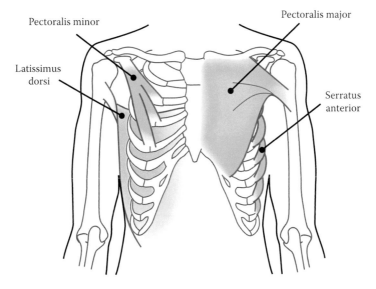

FIGURE 3.8
Normal anatomy of key musculature for shoulder disarticulation TMR surgery.

FIGURE 3.9
Plane between the clavicular (left) and sternal (right) heads of the pectoralis major muscle. The brachial plexus lies within the fatty tissue exposed by this dissection.

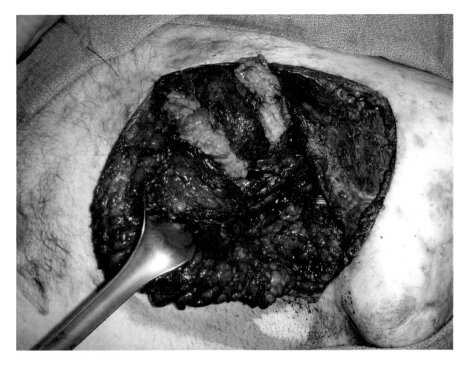

FIGURE 3.10
Medially based adipofascial flaps positioned between the clavicular head and two sternal segments of the pectoralis major muscle.

FIGURE 4.1
Median nerve end-neuroma encountered during a transhumeral TMR procedure.

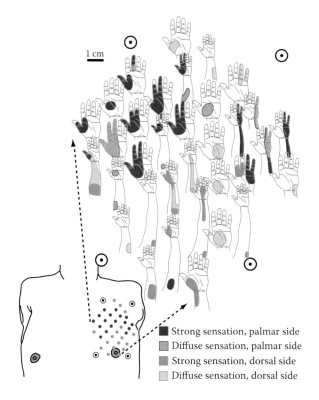

FIGURE 8.1
The reinnervated chest skin of a shoulder disarticulation TMR patient showing sensations referred to the missing limb elicited by indentation of the reinnervated skin by a cotton-tipped probe (300 g applied force). (From Kuiken et al., *Proc Natl Acad Sci U S A* 104 (50):20061–20066, 2007. With permission.)

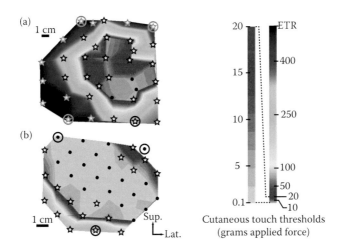

FIGURE 8.2
Contour plots showing the average amount of force required for two patients (a and b) to feel touch to the reinnervated chest projected to the missing limb. Black dots indicate points where only the missing limb was felt. Stars indicate points where chest was felt at lower force thresholds and the missing limb was felt at higher thresholds. (From Kuiken et al., *Proc Natl Acad Sci U S A* 104 (50):20061–20066, 2007. With permission.)

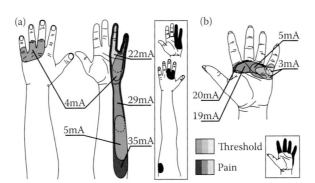

FIGURE 8.3
Projected sensations elicited by electrical stimulation in two targeted reinnervation amputees (a and b). Projected fields in blue indicate sensation (threshold) levels, and those in red indicate painful stimulation levels. Insets: Composites of mechanosensory projected fields (300 g applied force) at the same positions/points where the electrical stimulation electrodes were placed. (From Kuiken et al., *Proc Natl Acad Sci U S A* 104 (50):20061–20066, 2007. With permission.)

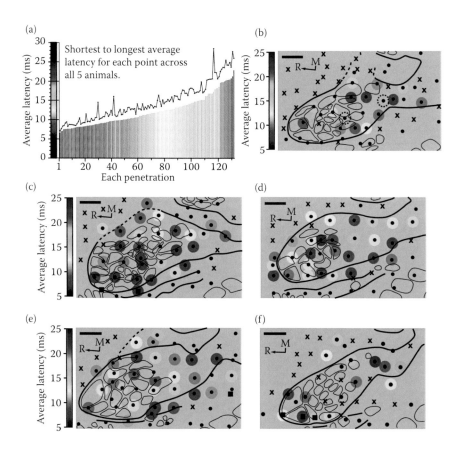

FIGURE 8.9
Latency measurements for electrically evoked somatosensory potentials from stimulus artifact to cortical response from the reinnervated target skin to the electrode penetration points
within the forelimb barrel subfield of five targeted reinnervation rats. (a) Average latency for
each penetration ($n = 132$) across all recording cases, arranged from shortest to longest. Black
line indicates SD for each latency. (b–f) Latency measurements for each electrode penetration
in the forelimb barrel subfield overlaid with cytochrome oxidase-delineated digit barrel structures. Colored circles indicate latency in milliseconds; (●) indicates a responsive site, (X) an
unresponsive site, (■) a dual receptive field site; R, rostral; M, medial. Dashed circles indicate
inability to measure average latency. For clarity, the diameter of the colored latency points is
set to reflect 1.5 times the 100-µm hypothetical recording sphere at the electrode tip (Robinson
1968). Average latencies for each penetration range from 6.6 ms to 23.0 ms. Often, shorter latencies lie directly adjacent to longer latencies. In (f), no latencies were recorded at four points
within the forelimb barrel subfield. Scale bars represent 500 µm. (From Marasco and Kuiken, *J
Neurosci*, 30 (47):16008–16014, 2010. With permission.)

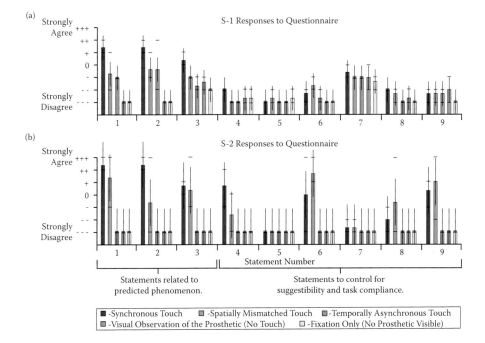

FIGURE 8.14

Questionnaire results for TMR subject S-1 (a) and TMR subject S-2 (b) for five different stimulus conditions. Vertical error bars indicate 95% confidence intervals (S-1 = ± 0.975, S-2 = ± 1.865) from a multiple comparisons procedure; horizontal lines indicate range, $n = 3$. Significance was judged by non-overlap of confidence intervals. Statements 1 through 3 were predicted phenomena: question #1, "I felt the touch of the investigator on the prosthetic hand"; 2, "It seemed as if the investigator caused the touch sensations that I was experiencing"; 3, "It felt as if the prosthetic hand was my hand." Statements 4 through 9 were controls: 4, "It felt as if as if my residual limb was moving towards the prosthetic hand"; 5, "It felt as if I had three arms"; 6, "I could sense the touch of the investigator somewhere between my residual limb and the prosthetic hand"; 7, "My residual limb began to feel rubbery"; 8, "It was almost as if I could see the prosthesis moving towards my residual limb"; 9, "The prosthesis started to change shape, color and appearance so that it started to visually resemble the residual limb." For S-2, confidence intervals for scores related to ownership of the limb did not overlap with control scores. There was some overlap between the ownership and control scores for S-1. (Adapted from Marasco et al., Robotic touch shifts perception of embodiment to a prosthesis in targeted reinnervation amputees, *Brain* 134 (Pt 3):747–758, 2011. With permission from Oxford University Press.)

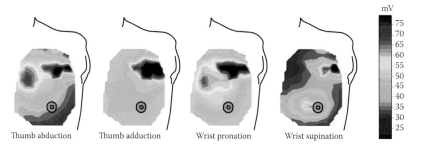

Thumb abduction   Thumb adduction   Wrist pronation   Wrist supination

FIGURE 10.1

Mean absolute values of the EMG signal amplitudes from 115 high-density electrodes generated when a TMR patient attempted the indicated test contractions.

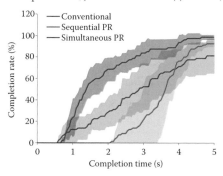

(a) Trial start   (b) Successful trial end

Virtual prosthesis

Target posture

FIGURE 10.4

A simple example of the TAC Test display presented to a user. The patient has real-time control over the virtual prosthesis and is required to move it to the target posture. In this example, the user is only required to perform wrist flexion to successfully complete the trial. (From Simon et al., Target Achievement Control Test: evaluating real-time myoelectric pattern-recognition control of multifunctional upper-limb prostheses, *J Rehabil Res Dev* 18 (6):619–628, 2011. With permission.)

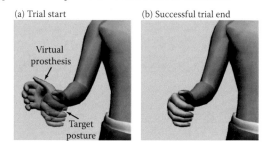

FIGURE 10.6

Functional performance results averaged across four TMR patients as they completed TAC tests that required simultaneous control of two degrees of freedom. Each subject completed the tests using conventional TMR control (Conventional), sequential pattern recognition control (Sequential PR), and simultaneous pattern recognition control (Simultaneous PR). Higher and faster completion rates are representative of better control. Shaded regions represent ± 1 SD. (From Young et al., A functional comparison of simultaneous pattern recognition to conventional myoelectric control, *J Neural Eng Rehabil.* In submission.)

5

Rehabilitation of the Targeted Muscle Reinnervation Patient

Todd A. Kuiken

CONTENTS

5.0 Introduction

Targeted muscle reinnervation (TMR) is an exciting new approach to improving prosthetic function for people who have had upper limb amputations. However, it requires a significant surgery, a long postsurgical reinnervation period, and a fairly intensive program of prosthetic fitting and training. The unique complexities associated with this process should be taken into careful consideration during both candidate selection and the rehabilitation of patients who have undergone TMR.

5.1 Presurgical Evaluation

5.1.1 Physical Examination

A thorough presurgical examination is important when considering TMR for any patient. A standard medical exam should be performed, with specific attention given to residual limb length, remaining musculature, and the health of the residual nerves. If the residual limb is long enough for the patient to be fit in a transhumeral socket and both biceps and triceps muscles remain, then the transhumeral TMR procedure is the clear choice. The shoulder disarticulation TMR procedure is appropriate for patients with true shoulder disarticulations and those with short transhumeral amputations that do not allow adequate control of a transhumeral prosthesis (i.e., patients with only a few centimeters of remaining humerus) (Figure 5.1).

Brachial plexopathy is a contraindication to TMR surgery, as healthy nerves are needed for successful reinnervation of target muscles and good shoulder support is needed for prosthesis use. Brachial plexopathy can be missed in an amputee because of the absence of distal musculature. In the transhumeral amputee, the residual limb must be carefully examined for adequate activation of the residual biceps and triceps, while the proximal shoulder should be thoroughly assessed for muscle activation, muscle bulk, and strength. Similarly, adequate muscle activation and bulk are expected in

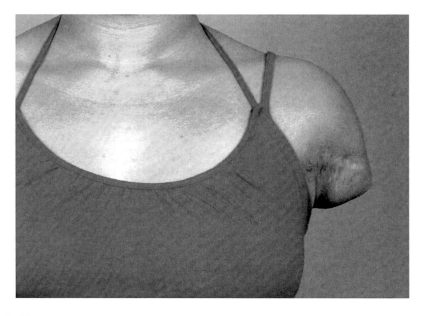

FIGURE 5.1
This patient with a very short transhumeral amputation was not able to control a transhumeral prosthesis and was therefore treated with the shoulder disarticulation TMR procedure.

patients with shoulder disarticulation: any undue atrophy of the pectoralis, supraspinatus, or infraspinatus muscles is reason for concern. Sensory testing should also be performed, but it is of less value as proximal sensation can be normal even with significant brachial plexopathy. Probing for positive Tinel's signs for the median, ulnar, and radial nerves is useful to determine the likely location of an end-neuroma. However, positive Tinel's signs do not rule out brachial plexopathy, nor does the lack of Tinel's sign indicate that the nerve is nonfunctional, as the neuroma may just be proximal and well protected by soft tissues. A diagnostic electromyogram (EMG) may be helpful if brachial plexopathy is suspected. The EMG will still show signs of denervation in any residual muscle, as it is no longer normal muscle: it has been damaged and will display spontaneous activity. However, the proximal shoulder muscle and paraspinal muscle should be normal. Magnetic resonance imaging (MRI) of the brachial plexus may have value in evaluating a patient for a brachial plexus injury such as nerve root avulsion, nerve discontinuity, or nerve atrophy. MRI can also be useful for locating end-neuromas and determining nerve length in preoperative planning, especially for patients with shoulder disarticulation.

The patient should be carefully examined for any additional injuries or conditions that may interfere with rehabilitation, prosthetic fitting, or prosthesis control. Joint injury proximal to the amputation site is common with traumatic amputation; therefore, the shoulder complex of transhumeral amputees must be examined for injury and instability. For unilateral amputees, close attention should be paid to the remaining arm. Any impairments of the sound limb can be magnified as the patient will use this limb to compensate for the lost limb. Furthermore, overuse syndromes of the sound limb are common with time. Core strength and posture can also be affected by upper limb amputation, leading to problematic asymmetry of the spine and impairment of cervical function.

Soft tissue defects, difficult bony prominences, split-thickness skin grafts, and challenging scars that can interfere with prosthetic fitting are also common after traumatic amputation (Figure 5.2); a thoughtful examination should be made to consider the possibility of correcting these problems when performing TMR surgery. Special note should also be taken of biceps contraction in the transhumeral amputee. Sometimes the distal biceps is not, and cannot be, adequately secured to the distal humerus. Thus biceps activation causes a contraction that significantly displaces the muscle in the proximal direction (Figure 5.3). If the biceps pulls far enough, it can be difficult or impossible for the prosthetist to acquire an adequate EMG signal. In such cases, cutting the proximal tendons of the biceps heads should be considered to prevent the biceps from retracting; this approach can drop the biceps 2 cm or more and greatly facilitate prosthetic fitting.

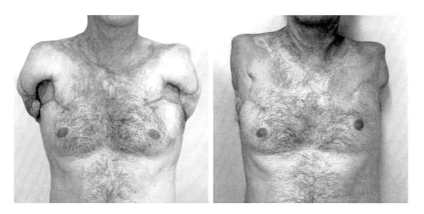

FIGURE 5.2
Left: Painful split-thickness skin grafts after shoulder disarticulation amputation. (Reprinted from Lipschutz et al., Shoulder disarticulation externally powered prosthetic fitting following targeted muscle reinnervation for improved myoelectric control, *J Prosthet Orthot* 18 (2):28–34, 2006. With permission.) *Right*: Skin grafts were revised during TMR surgery.

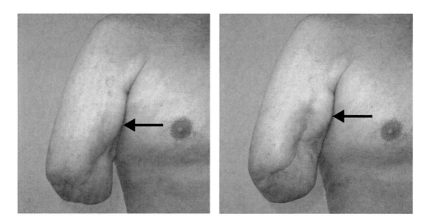

FIGURE 5.3
Left: The biceps muscle in a patient with a transhumeral amputation (center of muscle belly indicated by arrow). *Right*: Contraction caused the biceps to migrate several centimeters proximally due to the loss of distal insertions. This motion can be problematic for prosthetic fitting and can be eliminated by cutting the proximal tendons to the muscle.

5.1.2 Cognitive and Emotional Evaluations

Cognitive and emotional evaluations are important parts of the initial evaluation. Cognitive or emotional impairments can greatly impede the patient's ability to participate in rehabilitation and to comply with treatment requirements.

Traumatic brain injury (TBI), while sometimes subtle, is a possibility with most traumatic amputations. For this reason, cognitive function screening should be performed and a workup should be pursued if there are any concerns. Mild TBI does not exclude a patient from having TMR surgery if the patient can still follow directions and comply with requirements; however, delaying TMR surgery until the patient recovers from a mild to moderate TBI is recommended.

Mental health issues are very common following amputation. The incidence of significant depression or anxiety following any limb amputation (mostly dysvascular amputation) has been reported as approximately 28% and 36%, respectively (Desmond 2007). Based on our experience, the incidence of depression is much higher in younger patients and in those who have experienced traumatic amputation. The distribution of depression onset after amputation is bimodal. It peaks first early after amputation and has a second peak 1 to 2 years later when the amputee, having finished rehabilitation, comes to realize that his or her condition will generally not improve any further (Rybarczyk et al. 1997). Individuals who undergo a traumatic amputation are also at high risk for developing post-traumatic stress disorder (PTSD) (Copuroglu et al. 2010). Even with nontraumatic amputations, the associated stress commonly results in anxiety disorders. Therefore, depression must be looked for carefully and continually so that appropriate treatment can be initiated when needed. Frank and repeated questioning about the patient's mood, sleep patterns, appetite, frustration level, and symptoms of anxiety are warranted. The entire clinical team should continuously monitor the emotional well-being of these patients, although usually the prosthetists and therapists spend the most time with patients and therefore gain better insight into their emotional state. We recommend that all amputees have an opportunity to talk with a psychologist as an investment in their emotional health. Depression and anxiety disorders do not exclude a patient from having TMR surgery, but these disorders must be controlled first, as TMR surgery and the rehabilitation process can cause significant additional stress that may exacerbate preexisting conditions.

5.1.3 Assessing Commitment and Compliance

The final part of the evaluation is a consideration of the patient's commitment to using a prosthesis and the patient's understanding of what TMR will and will not do. This is necessary to determine the likelihood of the patient's compliance with treatment. Typically, a patient will have already been fit with a prosthesis for daily use prior to considering TMR. If the patient continues to use this prosthesis at least occasionally, the improved control offered by TMR should be considered a valuable addition to an already acceptable device. However, if the patient has completely abandoned the use of the prosthesis, then extra care must be taken to ensure that he or she understands the relative value of TMR: it must be made clear that the TMR prosthesis will not be any lighter and will likely not fit any better or be any more comfortable than the previous prosthesis. In light of these facts, the patient must determine if the improved control will be enough to result in regular wear.

Finally, attention should be paid to factors that may indicate the patient's likelihood of compliance with treatment. If patients have a history of not making or keeping appropriate follow-up appointments with their prosthetist to maintain their device in adequate working condition, or if they were similarly unreliable in previous therapy, then future noncompliance should be suspected and these patients should be dissuaded from having TMR surgery.

5.2 Patient and Team Education

Patients must be well informed of the details of their individual TMR procedures, expectations for recovery, and potential complications. It is important that all members of the rehabilitation team clearly understand which nerve transfers are being performed. Also important is that the team works together to develop a comprehensive treatment plan and schedule for these patients and their families.

After TMR surgery, the first twitches of reinnervation are very small and usually occur between 8 and 12 weeks after surgery. Muscle recovery then evolves more rapidly: fairly strong palpable and recordable contractions can be elicited in just another month or two. This is exciting to the patient and the whole rehabilitation team; however, we discourage TMR prosthetic fitting until at least 6 months after surgery. Successful control cannot be obtained with an early fitting, because reinnervation and muscle strengthening are not complete. The best EMG sites for a given arm movement may change location. Therefore, fitting the patient too early generally results in the entire clinical team attempting to "chase" changing EMG signals, which is time consuming, unproductive, and very frustrating for both the team and the patient. Furthermore, the patient and team must realize that proximal nerve-to-nerve coaptations will take even longer to fully establish stable, functional connections with target muscle. Thus, while some muscles may reinnervate in the predicted time frame, reinnervation of others may be delayed and such delays should be anticipated by the patient and the clinical team. Finally, before prosthetic fitting is initiated, we recommend that the surgeon meet with the patient and rehabilitation team to review the surgery, emphasizing which muscle contractions are expected with which attempted movements and clarifying any recovery or prosthetic fitting issues. This is especially important for shoulder disarticulation TMR patients, as the surgical procedure is more complicated and has quite variable outcomes in these individuals.

5.3 Potential Surgical Complications

TMR is a soft tissue surgery with the same risk of complications for infection, hematoma, and general anesthesia as other soft tissue surgeries. However, there are a few potential complications that are specific to TMR and that should be discussed with the patient.

5.3.1 Risk of Muscle Paralysis

There is some risk that the muscle paralysis caused by cutting the native innervation of target muscles during surgery will be permanent, despite subsequent nerve transfer. However, this risk is small because the large brachial plexus nerves contain many times the number of motor axons needed to fully reinnervate the muscle (see Chapter 2 for a discussion of hyper-reinnervation of target muscles). The slight risk of permanent paralysis is tempered by the fact that muscles used in TMR are generally not functional; therefore, permanent paralysis of these muscles will not further reduce functional performance.

5.3.2 Neuroma Pain

Any cut nerve has the potential to develop a painful end-neuroma. The creation of end-neuromas is therefore a risk with TMR surgery. However, TMR may be an effective treatment for painful neuromas resulting from amputation; during TMR surgery, existing neuromas are resected and nerves are allowed to reinnervate a denervated target muscle (see Chapter 4 for details on neuroma formation and treatment).

5.3.3 Phantom Limb Pain

After TMR surgery, patients generally experience a transient increase in their phantom limb pain, followed by a return to baseline pain levels. The phantom pain experienced after TMR is usually not as severe as the pain experienced following the initial amputation, likely because there is little accompanying trauma and the patient is in a much better emotional state. The increased phantom limb pain can be reduced with initiation of neuropathic pain medication or an increase in the dosage of current medication. The increase in phantom limb pain generally lasts from 4 to 6 weeks.

5.3.4 Transfer Sensation

The nerves of the brachial plexus have a mix of motor and sensory fibers. When these nerves are transferred to denervated muscle segments, the sensory nerve fibers regenerate through the muscle and reinnervate any denervated skin (Kuiken et al. 2007). When this skin is touched, the patient feels as if the missing hand has been touched (a phenomenon termed *transfer sensation;* see Chapter 8 for further discussion). In general, little or no transfer sensation occurs unless the sensory nerve that innervates the skin overlying the target muscle is purposefully cut. However, small areas of transfer sensation will sometimes develop without previous denervation of the skin. The sensation generated by touching reinnervated skin may feel like pressure or a dysesthesia (tingling feeling) on the missing hand. Patients must be informed prior to surgery that this may occur. By and large, patients have not been concerned when experiencing this sensation, and many even enjoy being able to feel part of their lost hand again.

5.4 Conclusion

Performing TMR requires effective communication between the patient and members of the surgical and rehabilitative teams. Before surgery, individuals contemplating TMR should be thoroughly evaluated to ensure that the procedure is possible and likely to be successful, that the individual is able to commit the effort and time necessary for rehabilitation, and that the procedure will provide sufficient benefit to the user. Individuals should be aware of the relatively minor risks associated with this procedure and the relatively long time required for reinnervation to evolve, as well as the potential benefits. For most individuals to date, TMR has provided significant benefits—including improved control and reduced neuromatous pain. Careful screening to ensure that individuals are appropriate candidates for the procedure, and careful coordination between patient and clinical team members, will ensure that TMR continues to improve the quality of life and functional ability of individuals who have had upper limb amputations.

References

Copuroglu, C., M. Ozcan, B. Yilmaz, Y. Gorgulu, E. Abay, and E. Yalniz. 2010. Acute stress disorder and post-traumatic stress disorder following traumatic amputation. *Acta Orthop Belg* 76 (1):90–93.

Desmond, D. M. 2007. Coping, affective distress, and psychosocial adjustment among people with traumatic upper limb amputations. *J Psychosom Res* 62 (1):15–21.

Kuiken, T. A., P. D. Marasco, B. A. Lock, R. N. Harden, and J. P. A. Dewald. 2007. Redirection of cutaneous sensation from the hand to the chest skin of human amputees with targeted reinnervation. *Proc Natl Acad Sci U S A* 104 (50):20061–20066.

Lipschutz, R. D., T. A. Kuiken, L. A. Miller, G. A. Dumanian, and K. A. Stubblefield. 2006. Shoulder disarticulation externally powered prosthetic fitting following targeted muscle reinnervation for improved myoelectric control. *J Prosthet Orthot* 18 (2):28–34.

Rybarczyk, B., J. J. Nicholas, and D. L. Nyenhuis. 1997. Coping with a leg amputation: integrating research and clinical practice. *Rehabil Psychol* 42 (3):241–256.

6

Prosthetic Fitting before and after Targeted Muscle Reinnervation

Laura A. Miller and Robert D. Lipschutz

CONTENTS

6.0 Introduction

Targeted muscle reinnervation (TMR) surgery provides additional myo-electric control signals for more intuitive prosthesis control and can result in markedly improved function; however, prosthetic fitting of TMR patients is challenging. It is important to tailor the prosthetic device to the individual both before and after targeted reinnervation surgery. Proper component selection and socket design maximize functional performance and allow the user to take full advantage of the benefits of TMR. Successful fittings require an experienced prosthetist with skill in myosite selection and extensive knowledge of myoelectric components. The prosthetic treatment of TMR patients is an iterative process that requires close collaboration between the prosthetist, physiatrist, occupational therapist, and surgical team.

In this chapter, we focus on the prosthetic fitting of individuals with transhumeral and shoulder disarticulation amputations at three different stages: prior to TMR surgery, in the postsurgical interim period during reinnervation, and after reinnervation. We review concepts for presurgical and postreinnervation control strategies and discuss relevant aspects of socket design and fitting. We also present principles of hardware and software selection for the TMR patient and outline the steps necessary for successful patient education and training.

6.1 Presurgical Considerations

6.1.1 Review of Surgical Goals and Requirements

6.1.1.1 Transhumeral Amputation

The desired outcome of TMR for transhumeral-level amputations is to obtain a minimum of four independent sites for intuitive control of elbow flexion, elbow extension, hand close, and hand open. Native innervation of the long heads of the biceps and triceps allows for myoelectric control of elbow flexion and extension, respectively. The short head of the biceps (reinnervated by the median nerve) and the lateral head of the triceps (reinnervated by the distal radial nerve) are used for hand-close and hand-open functions, respectively. Ulnar nerve transfer to the brachialis, if performed, can be used to control a wrist rotator or as an input for hand control if the median or distal radial signals are too weak.

In general, TMR candidates with transhumeral amputation should have, at minimum, a mid-length residual limb to provide an adequate lever arm. Transhumeral limbs that meet the length criteria of *standard transhumeral level*—approximately 50% to 90% of original limb length—are the most appropriate; this optimal limb length permits an externally powered elbow to be used without causing segmental length discrepancies. Shorter residual limbs may lack sufficient residual muscle to provide adequate muscle targets. Longer residual limbs may be suitable for these fittings based on the mechanical advantage of the longer lever arm, provided that the individual can accept the resulting non-anthropomorphic segment length. Specific surgical techniques that address fitting issues—such as an angulation osteotomy (see Chapter 3)—may add significant benefit for an individual with a longer residual limb.

6.1.1.2 Shoulder Disarticulation Amputation

The desired outcome of TMR for the shoulder disarticulation amputee is also to obtain at least four control sites for elbow flexion, elbow extension, hand close, and hand open. However, the surgical plan for the shoulder disarticulation level may vary greatly depending on the remaining musculature, nerve locations, and anatomic structures of each patient. Possible target muscles for individuals with shoulder disarticulation amputation include segments of the pectoralis major, pectoralis minor, latissimus dorsi, and serratus anterior. Reinnervation of the latter two target muscles is achieved by coaptation of transferred nerves to the long thoracic and thoracodorsal nerves, respectively. Because transferred nerves must regenerate over a longer distance to reach these target muscles, prosthetic fitting will be correspondingly delayed.

6.1.2 Candidate Selection

Candidates for TMR surgery should be thoroughly evaluated by the entire clinical team. To be successful with TMR, patients must have a suitable residual limb(s), a psychological profile indicating that they will likely be compliant and patient with the demands of the process, and an appropriate understanding of the demands and benefits of the TMR procedure. The essential elements of physical and psychological evaluations for prospective TMR patients are detailed in Chapters 3 and 5.

The prosthetist should perform a thorough physical exam with special regard to prosthetic fitting issues. In particular, any soft tissue or bone problems or other issues that may result in a suboptimal fitting should be noted. Many patients have excessive soft tissue (mainly fat) in the upper arm and axillary regions; transhumeral patients may have excess soft tissue in the distal residual limb. Scars or abrupt changes in tissue contours can also be problematic for optimal socket fitting. These issues should be discussed with the surgical team, as it may be possible to remove excess soft tissue, smooth out tissue transitions, or revise the soft tissue envelope to provide greater coverage of bony prominences during TMR surgery. Or, if the location of the biceps control site is too proximal, the surgeon may be able to cut the proximal tendons of the biceps, thus lowering the muscle belly 1 to 2 cm (see Chapter 5, Figure 5.3). The key point here is that the prosthetist will appreciate and anticipate fitting issues better than the surgeon will. Good communication within the team is essential to allow for the discussion and optimal resolution of such problems.

Each individual should be evaluated as a candidate for using myoelectric prosthesis control. As for any myoelectric fitting, a basic understanding of the abilities and limitations of myoelectric devices is required, including the need to charge the battery regularly, use components appropriately, and understand the control strategy and method of mode selection. If the candidate currently uses a myoelectric prosthesis, it should be made clear that only control of the device will change after TMR as, in most cases, the socket and components will remain essentially the same.

When contemplating TMR surgery, candidates must understand the basics of the procedure, the benefits of the surgery, the training and therapy that will be required to attain those benefits, and the time commitment required for a successful prosthetic outcome. Appropriate expectations of TMR surgery are important. If a myoelectric system has previously been tried and rejected because of weight or inadequacy of component function, it is unlikely that the candidate will accept a similar prosthesis after TMR surgery. However, TMR may be appropriate if an individual feels that improved control will make the prosthesis more beneficial. It is important that the patient be emotionally stable and capable of cooperating through the lengthy process of TMR surgery, reinnervation, prosthetic fitting, and training. The prosthetist often spends more time with the patient than any other member of the clinical team, especially early in the process, and thus develops unique insights about the patient. Any concerns about the patient's mood, ability to comply with the treatment program, or other psychosocial issues should be discussed with the patient's physician.

6.1.3 Presurgical Fitting

Once a suitable candidate has been identified, the patient should be evaluated for presurgical fitting. There are two scenarios for prosthetic fitting prior to planned TMR: (1) the patient with a recent amputation and a well-healed residuum who wishes to undergo initial prosthesis fitting prior to surgery and (2) an individual who elects to undergo TMR several years after an amputation.

With a recent amputee, fitting with a traditional myoelectric or body-powered prosthesis prior to TMR surgery can be useful. Early prosthetic fitting allows the individual to get used to the weight and function of a prosthesis and to incorporate its use into daily activities. Studies have shown that individuals will continue to use a prosthesis more frequently if fitted soon after amputation (Malone et al. 1984; Pezzin et al. 2004). In addition, if the individual is trained to use a prosthesis and to incorporate it into bimanual activities before surgery, then, after surgery, the individual can focus on adapting to control changes rather than having to simultaneously learn both control and function. If the patient is receptive to a body-powered device, this is generally the easiest method for providing a useful prosthesis before surgery and during the interim reinnervation period. It can then serve as an alternative device for the patient after TMR fitting. Socket modifications may be required following surgery, but the same cable-driven control strategy can be used until reinnervation has occurred.

If a new myoelectric device is fitted prior to surgery, the prosthesis should be capable of using four or more electromyographic (EMG) input channels after TMR. Some commercial arm systems only support two EMG inputs or are difficult to use with more than two EMG sites. Additionally, using remote (dome style) electrodes (Figure 6.1a) rather than packaged electrodes (Figure 6.1b) provides more flexibility if modifications to electrode position are necessary following surgery. When fabricating a presurgical socket for a

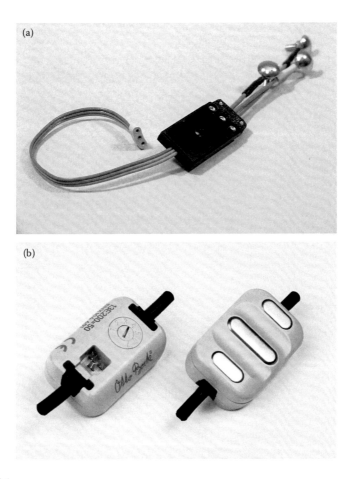

FIGURE 6.1
(a) Remote electrodes (i.e., contacts that are separate from the amplification and filtering circuitry). (LTI Remote Electrode System, photograph provided courtesy of Liberating Technologies, Inc.) (b) Packaged electrodes (i.e., contacts that are "packaged" with the amplification and filtering circuitry) (MYOBOCK® Electrodes, Otto Bock HealthCare GmbH, Duderstadt, Germany, photograph provided courtesy of Otto Bock.)

transhumeral amputee, electrodes should be placed over the long head of the biceps and the long head of the triceps as these muscles will not be affected by surgery. Socket modifications after surgery will generally be minimal. If TMR surgery is planned within a short time frame, a prolonged test socket may be used instead of a definitive socket. If a final socket is constructed prior to surgery, a flexible inner socket and frame is recommended so that any significant changes in limb volume or electrode location after surgery can be accommodated by replacing only the inner socket. Extra space should be allowed between the flexible liner and frame for the additional electrodes and wires that will be needed after reinnervation. The best choice of control strategy or method of mode selection to be used prior to TMR is open to debate. To date, there is no evidence that one initial strategy for pre-TMR prosthesis fitting is more effective than another with respect to functional control after TMR.

For transhumeral TMR candidates with a long-standing amputation, presurgical fitting strategies depend on their previous experience with a prosthetic device. Nothing needs to be done initially if they are successful prosthesis users with well-fitting devices. If the prosthesis is not working or is ill fitting, then the prosthetist should determine if it can be readily repaired and adjusted or if it is better to simply wait until after TMR surgery and reinnervation.

At the shoulder disarticulation level, most available muscles will be used as target sites and therefore will be denervated during surgery. For unilateral amputees, it may not be worthwhile to make a new presurgical or interim prosthesis, as control of these devices may be difficult to master and will change dramatically after reinnervation. Having to learn two new control strategies within a relatively short time frame may be too challenging and frustrating for these patients. In these cases, we generally recommend that prosthesis use be deferred until the final TMR fitting, even though, as previously mentioned, the patient may be receptive to, or continue to use, a body-powered device. For bilateral amputees, interim devices are necessary and at least one body-powered arm is recommended. Again, for powered devices, EMG control should be avoided in order to minimize the need for significant postsurgical retraining. Switches, force-sensing resistors (FSRs), and/or linear transducers should be used as inputs instead of myoelectric control.

6.2 Postsurgical Reinnervation (Interim) Period

6.2.1 Prosthetic Fitting Considerations

After TMR surgery, it generally takes 3 to 6 weeks for the surgical edema to abate and for the limb shape to stabilize. At the transhumeral level, it is likely that adipose tissue will be removed from, or relocated within, the residual limb during TMR surgery and that residual limb volume will change as a consequence. If the changes are subtle, the socket can be modified by inserting padding between the flexible inner socket and frame. If a duplicate of the socket was kept, it may be possible to modify the volume by creating a new flexible insert without needing to remake the frame. When the volume change is substantial, resulting in excessive room in the socket and/or inadequate skin-electrode contact, a replacement socket is warranted.

For myoelectric devices, it may be necessary to alter electrode locations to accommodate any shift of the muscle bodies during surgery. This may be accomplished quite easily if remote electrodes were used in the presurgical prosthesis. It may also be necessary to increase electrode gains (boost) to accommodate the decrease in active muscle mass following surgery due to denervation of some muscle segments.

At the shoulder disarticulation level, depending on any necessary changes to the control input and the amount of soft tissue reconstruction, the user should be able to wear the presurgical socket during the interim reinnervation period with little or no modification.

6.2.2 Control Strategies during Reinnervation

During reinnervation, transhumeral amputees can continue to use the same control strategy as before surgery. Similarly, shoulder disarticulation patients should require minimal modification to their prosthesis if they are using body-powered devices or motorized devices controlled with switches, FSRs, or other mechanical means. However, myoelectric prostheses will not operate adequately if the control strategy relied on EMG from muscles that were denervated during surgery. For this reason, as noted earlier, it may be wise to defer fabricating a prosthesis for unilateral shoulder disarticulation patients until reinnervation is complete. For bilateral amputees, it is important to modify their devices so that they use mechanical controls during this interim reinnervation period.

6.2.3 Postsurgical Education and Training

Patients should be encouraged to follow standard postsurgical care, including wound dressing and medication for pain management. Patients should not wear a prosthesis for 3 to 6 weeks to allow surgical healing. Immediately after surgery, the limb will be numb if skin was denervated through removal of subcutaneous fat or through deliberate division of sensory afferents to allow for sensory reinnervation, and care should be taken to protect these areas. Individuals should be reminded to exercise to maintain range of motion, to perform massage for desensitization, and to wear a shrinker or rigid dressing as appropriate.

If the individual has not used a device for an extended period of time or if the control strategy has been changed following surgery, a few postsurgical sessions of occupational therapy might be helpful. If possible, use of a device during the reinnervation period should be encouraged.

6.3 Post-Reinnervation Fitting

6.3.1 Myotesting

Although it is usually possible to record EMG signals from target muscle sites about 3 months after TMR surgery, the signals will be weak and will continue to change since reinnervation is still actively taking place. Attempting to identify control sites too early can be frustrating for both the patient and the clinical team. The most efficient strategy is to delay control site selection until reinnervation begins to stabilize at around 6 months after surgery. If signals are weak, the individual can work with an occupational therapist to put together a home program of exercises to increase muscle endurance and signal magnitude, which will make control site selection easier (see Chapter 7 for a recommended exercise program).

The operative report can be very helpful for understanding which nerves were transferred to which muscles. However, if at all possible, an examination of the patient in conjunction with the surgeon is highly recommended early in the myotesting phase. This can be very helpful in assessing the final surgical outcome. Getting the patient to perform different hand or arm functions can clarify the final reinnervation pattern, which may be quite different than what would be predicted from the operative report and can greatly focus and facilitate myotesting.

Myotesting begins with locating areas of muscle response for each intended arm movement. Although there are more than two available control sites after TMR, myotesting can still be performed using a two-site myotester (Figure 6.2). The clinician should describe actions rather than muscle contractions for required attempted movements (i.e., "bend elbow" instead of "contract biceps"). It may be helpful for the individual to mirror the attempted movement with the contralateral limb, if possible. The individual's ability to perform isolated contractions and to co-contract should be evaluated, as co-contraction of an antagonistic pair of muscles may be desirable as an input for mode selection. Control will also be enhanced by the user's ability to vary the signal strength, resulting in a corresponding change in output speed. This *proportional control* is beneficial for specific tasks such as grasping small and delicate objects. Optimal orientation of electrodes is necessary in order to detect EMG of sufficient magnitude for prosthesis control. The prosthetist should keep in mind that the amputation procedure, TMR surgery, or

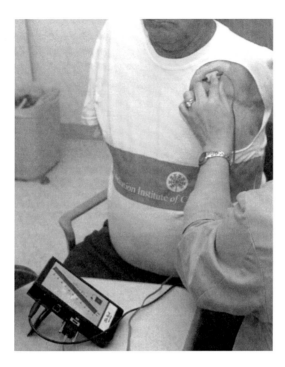

FIGURE 6.2
A prosthetist uses a two-site myotester to evaluate EMG sites in a shoulder disarticulation fitting after TMR.

subsequent surgical closures may have resulted in nonphysiologic muscle orientation. Optimal electrode placements may be determined by palpation; however, a trial-and-error approach is often required.

If the skin overlying the target muscles was also reinnervated, touch sensitivity may be more acute than before surgery (see Chapter 8 for a description of sensory reinnervation). The limb should therefore be palpated carefully to avoid causing discomfort.

It is important that the prosthetist use all available information to ensure proper and efficient localization of control sites. A thorough working knowledge of upper limb anatomy is necessary, and a good anatomy reference is helpful at this stage. A book on weight lifting can be used to demonstrate to the patient that certain movements isolate particular muscle groups. Although it is helpful to have notes on the surgical procedure, this should be used only as a guide, as it is not possible to predict which functions will be represented in the hyper-reinnervated muscles and it is possible that nerves were incorrectly identified during the TMR procedure. It is therefore important to examine all movements associated with each transferred nerve when evaluating target muscle response. For example, when evaluating reinnervation by the median nerve, it is helpful to have the individual try wrist pronation, wrist flexion, and hand close.

In addition to generating sufficiently strong EMG signals, contractions of agonist/antagonist muscle pairs must be adequately *isolated*—that is, a signal can be generated from the required muscle while the signal from the antagonist (or other adjacent muscles) is relatively low. The best approach to obtaining isolated signals is to first locate areas that provide an adequate EMG signal for each movement and then progress through the iterative process of checking for signal isolation with respect to the signals from all other reinnervated (and natively innervated) control sites. Isolated signals from antagonistic muscle pairs provide optimal intuitive control; however, complete isolation of these signals is rarely possible. Antagonist muscles often work in tandem to stabilize joint segments or to control the rate at which a movement is being performed. Some movements, such as elbow flexion and hand close, are often synergistically associated, which can cause unwanted co-contraction. In addition, after TMR, many potential control sites are located within a relatively small muscle volume, making physical separation of these signals more challenging. It is incumbent on the user and clinical team to devise an effective control strategy given this typical, less than ideal situation. Because of difficulty in achieving EMG signal isolation, it may not be possible to use the strongest EMG signals for each intended movement. In most cases, electrode gain and thresholds need to be adjusted to achieve separate signals. Having access to multiple myotesters is useful at this stage, although not required.

During this process, the prosthetist must keep in mind that a fundamental goal of TMR is to allow the individual to control each prosthesis movement using an analogous attempted movement of the missing limb. Attempted movements that are well beyond being physiologically relevant (e.g., extending the "pinkie" [fifth finger] to control elbow flexion) should be avoided. As reinnervation progresses, different attempted movements may result in better control signals. For example, the best hand-close signal may initially be obtained by attempting wrist flexion, although this is not optimal because the attempted movement is not directly related to the intended prosthesis movement. However, as reinnervation continues and the user continues to practice, it may be possible to use a more physiologically appropriate attempted movement, for example, attempting to flex the index finger or to close the hand. Often, time and practice are required to obtain the best congruency between attempted arm movement and desired prosthetic function. Even after the 6-month reinnervation period, EMG sites selected during a fitting should be evaluated at subsequent fittings, as reinnervation will continue to evolve and different attempted movements may generate better signals and/or optimal control site locations may change. As an example, our first TMR subject was recently refitted with a new TMR prosthesis. In the 5-year interval between this fitting and his previous fitting, the location of the "best" signal for some control sites shifted by as much as 2 inches.

Target muscles denervated during TMR surgery will be paralyzed until reinnervation takes place and, as a result, may atrophy following surgery. Thus reinnervated muscle will initially become fatigued after a relatively short duration of activity; signal strength will decrease if the individual is unable to sustain a contraction due to fatigue. Natively innervated muscles in the residual limb may also atrophy as a result of disuse, although not to the same extent as the denervated muscles. Thus a good signal found early in a myotesting session that becomes weaker or disappears during the course of the testing session likely indicates muscle fatigue. Individuals may also begin to contract all muscles harder to compensate for fatigue, resulting in an increase in co-contraction across all signals. Initial myotesting sessions should not last more than an hour at a time, although a few sessions in a day may be possible if adequate rest periods are provided. In addition, signals from reinnervated muscles will initially be much weaker than those from muscles with intact native innervation. Individuals may need to contract natively innervated muscles more gently so that their EMG signals do not overwhelm those from reinnervated muscles.

6.3.2 Progression from Myotesting to Prosthetic Fitting

It is helpful to move to a system that provides feedback to the user as soon as possible. If available, a computer-generated virtual arm that responds to all control sites can be used for training. This allows multiple control options for a particular movement to be evaluated. By providing visual feedback of arm function, practice with a virtual system can help the user work on reducing unwanted co-contraction and synergistic contractions without the additional challenges of wearing the socket and actually moving the prosthetic arm.

If the prosthetic components are available, it is also helpful to practice controlling a prosthesis that is resting on, or clamped to, a table. In addition to allowing the user to operate the prosthetic system without having to wear it, this allows the controller and electrodes that will be used in the final system to be tested and initial thresholds and electrode gains to be established. If packaged electrodes are used, they can be taped to the skin or held in place with an elastic stockinet. Remote electrodes may be embedded in plastic and held in place in the same manner. An alternative is to solder size 3 fabric store snaps to the ring terminals on the wire harness. These can then be "snapped" onto self-adhesive electrodes, such as Noraxon Dual electrodes (Noraxon USA Inc.) or GS26 electrodes (Bio-Medical Instruments Inc.). The fabric snaps are removed when fitting the socket. Figure 6.3 shows self-adhesive electrodes being used to determine the ideal locations for EMG control sites during a myotesting session. All of these approaches allow electrodes to be moved around to evaluate different electrode locations, as desired.

Once suitable sites have been located, the electrodes can be moved into a check socket and the fit refined to maintain electrode contact as the user attempts various movements. After TMR, movement of skin due to

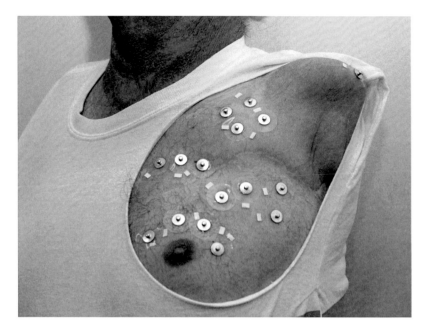

FIGURE 6.3
Self-adhesive electrodes used during myotesting after shoulder disarticulation TMR to evaluate potential control sites.

underlying muscle contraction is more dramatic than in the non-TMR case. It is likely that skin-electrode contact that is adequate when the user is not attempting movement will not be maintained during muscle contractions—this is particularly evident in the shoulder disarticulation case. The socket must therefore be modified so that electrode contact is maintained during both muscle relaxation and muscle contraction. Finally, the prosthetic components can be mounted onto the socket and both socket fit and control can be reevaluated with the full weight of the arm system in place.

6.3.3 Post-Reinnervation Fitting for Transhumeral Amputation

In the post-reinnervation TMR fitting, EMG signals from both natively innervated and reinnervated heads of the biceps and triceps are used to control elbow and hand function, respectively. The long head of the biceps muscle, with native innervation by the musculocutaneous nerve, is used for an intuitive elbow flexion signal. The long head of the triceps, with native innervation by the proximal radial nerve, provides an intuitive elbow extension signal. The short head of the biceps, which is reinnervated by the transferred median nerve, is used to close the hand, and the lateral head of the triceps, which is reinnervated by the distal branch of the radial nerve, is used to open the hand. It is often difficult to isolate signals for terminal device open and elbow extension because of cross talk—since the lateral and long heads of the triceps muscles are adjacent to each other—and because these movements are part of a synergistic "extension" pattern.

In a long transhumeral residual limb, the ulnar nerve can be transferred to the brachialis to provide a fifth myoelectric control site. The ulnar nerve controls some wrist flexion and both hand-open and hand-close functions of the intrinsic muscles of the hand. Thus an ulnar control site can be used in several ways. If either the median (hand-open) or radial (hand-close) nerve transfers do not provide robust control signals, then the ulnar control site can be used as a substitute. If all four other sites provide robust control signals, the ulnar site can be used for single-site wrist rotation. Although the ulnar nerve does not control wrist rotation in the intact arm, with time and practice the patient can learn to use an ulnar control site for this purpose. Although this may seem contradictory to the TMR philosophy of using physiologically appropriate control sites, this approach may prove to be more natural than other switching options.

Wrist rotation is the only other degree of freedom that is available for active control as currently available wrist flexion/extension units are all passive. If ulnar transfer is not performed, there are various options for controlling this action depending on the user's desire for simultaneous control of the wrist, ability to separate signals, and ability to contract or co-contract at variable rates. Simultaneous control of the elbow, wrist, and terminal device can be achieved using an electromechanical switch for wrist rotation. If the user has the ability and wishes to use a strategy that employs only EMG signals, the wrist and hand (or wrist and elbow) can be controlled sequentially with the same two EMG signals, using co-contraction or a rate-sensitive mode selection strategy. However, EMG signals from newly reinnervated muscles will initially be small. In addition, attempted co-contraction of the two hand-control sites frequently results in contraction of muscles that control the elbow, and the resulting inadvertent elbow movement may be frustrating for the user. It is also possible to begin with manual prepositioning of the wrist; a powered wrist can be added later, when the EMG signals are stronger and the patient is able to control the elbow and hand independently.

For the transhumeral fitting, a flexible inner liner and laminated frame are ideal because they allow replacement of the inner liner without completely remaking the device. This may be necessary for several reasons. Ongoing reinnervation and muscle strengthening may cause optimal control sites to shift over time, although this is minimal in comparison to that seen in shoulder disarticulation fittings. If packaged electrodes are used, a new inner socket will be necessary after reinnervation to accommodate additional electrodes. If remote electrodes are used in conjunction with a flexible inner liner, it is easy to move the electrode sites by making new holes and heat-sealing the old ones closed. Certain movements and contractions may affect socket fit, requiring modifications either to the socket or to electrode gains and thresholds.

Maintaining electrode-skin contact requires aggressive socket modifications. The short head of the biceps is the biggest challenge, as an intact proximal tendon will cause the muscle belly to contract proximally, often up to the anteromedial border of the socket. Electrode contacts must therefore be placed as proximally as possible without compromising either user comfort or the mechanical connection between electrode and socket. Improved user comfort is another reason to use remote electrode contacts. Another issue is that positioning the arm in space, especially far from the body, will apply torques to the socket that can cause electrodes to pull away from the skin surface. In a TMR fitting, this problem is exacerbated because there are more electrodes that must remain in contact with the skin. Positioning of the arm in space may also result in cross talk from deltoid EMG signals that interfere with target muscle EMG signals. Care should be taken when placing any electrodes (at natively innervated or reinnervated sites) to verify that cross talk from the shoulder musculature does not cause an inadvertent signal. It may be necessary to use a more distal control site, away from the deltoids, even if this means using weaker control signals.

6.3.4 Post-Reinnervation Fitting for Shoulder Disarticulation Amputation

Following successful reinnervation, multifunction prosthesis control by shoulder disarticulation amputees is achieved with a combination of EMG signals recorded from reinnervated muscles and natively innervated muscles along with mechanical inputs and postural controls. Muscle segments reinnervated by the musculocutaneous, radial, median, and ulnar nerves are generally used to control elbow flexion, elbow extension, hand close, and hand open, respectively. Often, remaining triceps or posterior deltoid can be used to control elbow extension, which allows the transferred radial nerve to be used for control of hand open. Additional functions, such as wrist rotation, can be added using a variety of methods.

Control site selection and maintenance of electrode-skin contact for a TMR fitting in a shoulder disarticulation patient presents more challenges than in a transhumeral fitting. Locating control sites that provide consistent, independent signals takes time and patience. A common target muscle for musculocutaneous nerve transfer is the clavicular head of the pectoralis major. The optimal control site is then located on the chest directly distal to the clavicle. Although this proximity to bone amplifies the signal—making it one of the strongest available—this proximity to the bony anatomy, and the potential for tissue movement caused by muscular contractions that cause movement at the sternoclavicular joint, makes socket fitting in this area difficult. The median nerve is frequently transferred to the upper half of the sternal head. Although this is a large, strong muscle, there is typically more subcutaneous tissue in this area to diffuse the EMG signal. Also, if the pectoralis is no longer inserted into the humeral head, contraction of the muscle will cause it to retract into a more medial location. This can also make it more difficult to identify the orientation of the muscle fibers during myotesting. The distal radial nerve is often transferred to the lower half of the sternal head of the pectoralis major, although it may be transferred to the long thoracic nerve (to reinnervate the serratus anterior)—usually done in female patients because the lower pectoralis major is covered with breast tissue. The pectoralis minor is frequently used as the target muscle for the ulnar nerve; this muscle may need to be moved laterally to position it out from under the pectoralis major. Finally, a nerve transfer to the thoracodorsal nerve (to reinnervate the latissimus dorsi) can be used when necessary. As noted previously, it takes a long time for a coapted nerve to regenerate down the thoracodorsal nerve or the long thoracic nerve to reinnervate the respective target muscles. Instead of the typical 3 months, it may take at least 6 months before initial EMG signals are detected; it may take several additional months for these muscles to be available as control sites.

If the distal radial nerve was transferred to the latissimus and the native triceps or posterior deltoids are used for elbow extension, separation of these two signals will take time and require user practice because of the close proximity of these muscles and their synergistic activation patterns. Also, signals from these sites are the most likely to be affected by postural movements.

EMG signals from reinnervated chest muscles may be very weak compared to the electrocardiograph (ECG) signal. In individuals with a left-side amputation, depending on the size of the individual and the location of the optimal EMG control sites, it may be necessary to use electrodes with modified filtering characteristics to better remove ECG signals. Liberating Technologies Inc. makes custom electrodes that have a high-pass filter at 60 Hz instead of a notch filter. These should eliminate much of the quiescent ECG signal, but the cardiac signal still might exceed EMG control thresholds when the user's heartbeat rises.

There are various options for control of the wrist. In one successful strategy, an FSR mounted in the socket was used to switch the myoelectric hand control signals to wrist control. In an alternate strategy, two socket-mounted FSRs were used to provide independent, bidirectional wrist control while allowing simultaneous EMG control of the terminal device. It is also possible to switch between control of the wrist or hand using rate-sensitive mode selection or co-contraction. However, these options will initially have similar problems as described for the transhumeral case (see Section 6.4.3) as a result of low initial EMG signal strength from reinnervated muscles. The range of motion of the remaining musculature should be also evaluated to see how shoulder movement might be used to actuate alternative switch inputs. These might be used to control a wrist rotator or to provide alternate hand and wrist control. If a powered shoulder lock is used, electromechanical switches are necessary to alternate between lock and unlock modes. Alternatively, the shoulder lock can be actuated with the contralateral hand

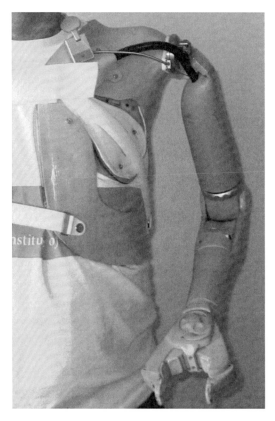

FIGURE 6.4
A modified Sauter-style socket with a flexible gel pad that ensures adequate electrode contact in a shoulder disarticulation TMR fitting.

or with a chin-nudge switch. If shoulder (acromion) or biscapular abduction is used to control the wrist, this movement may also influence the contraction of the target muscles and cause unwanted EMG signals. These movements can also alter the fit of the socket and cause electrode lift-off.

Given the diversity of potential EMG locations, no one socket design will be ideal in all cases. Trim lines for TMR fittings at the shoulder disarticulation level will often be dictated by control site location. Increasing the surface area of Micro-frame or X-frame designs (Miguelez and Miguelez 2003; Miguelez et al. 2004) may be necessary to encompass all of the optimal sites and to reduce the forces applied by the anterior strut, which, in combination with the domes of the electrode contacts, may exceed tissue tolerance. A Sauter-style frame socket (Figure 6.4) can also be used (Sauter 1992). If these low-profile frame designs provide insufficient coverage of the reinnervated area, externally mounted electrodes can be used (Figure 6.5). Because of the varied distribution of the EMG locations and the need to suspend the weight

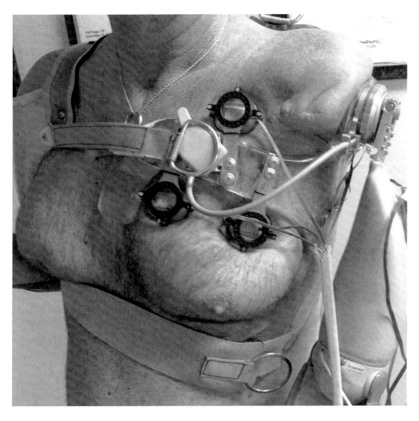

FIGURE 6.5
External spring-mounted electrodes. (Figure reproduced courtesy of Advanced Arm Dynamics.)

of an entire externally powered arm, starting with a larger socket is recommended. Once the electrode locations are finalized and the load-bearing areas identified, any unnecessary material can be removed to make the socket lighter and cooler. Cutouts may also be made to hold undergarments out of the way of the electrode contacts (Figure 6.6). A complete flexible inner socket allows increased loading with improved comfort but is heavier. Muscles on the chest wall may move with respect to the flexible inner socket during contraction, shoulder and torso movement, and with various arm positions or movements. To maintain electrode-skin contact, the flexible inner socket will need to be very tight. If weight is a significant issue, the inner socket can be eliminated and a flexible pad, made out of a thermo-formable plastic, can be used only in the areas of the electrodes (Figure 6.4). The addition of silicone to the flexible inner socket material may help protect areas of scarring or grafted skin. It is important to remember that room is required for the additional electrode preamplifiers and wires that have to be included between

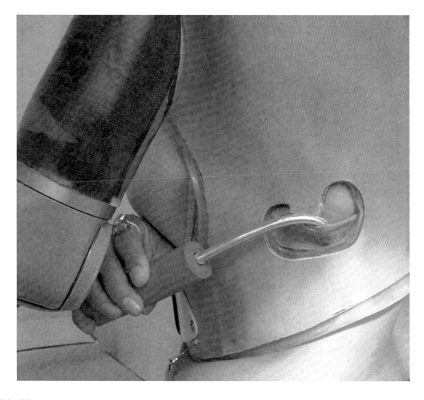

FIGURE 6.6
The ideal location for the "hand open" site in a female shoulder disarticulation patient with a nerve transfer to the serratus anterior was directly under the back strap of the brassiere. A cutout was made in the socket to hold the strap out of the way without having to modify the undergarment.

the flexible socket and frame. Spring compression devices can also be used to help hold packaged electrodes against the skin (Figure 6.5).

Positioning of the prosthesis and residual limb in space can cause unintended consequences. Given the torques that can be generated by an externally powered elbow, it is important to verify control during both active and passive use of the elbow with the shoulder in various locations. Postural effects may cause unintended signals. The following two clinical examples illustrate how such problems may be addressed:

- In a shoulder disarticulation fitting where the distal radial nerve was transferred to the long thoracic nerve/serratus anterior muscle, the ideal control site for hand open was along the side of the chest. When the user abducted the *contralateral* arm, cross talk from the oblique muscles (involved in postural alignment) generated a hand-open signal. Positioning the electrode in a more superior location reduced this interference, although the EMG control signal was weaker.

- A bilateral shoulder disarticulation amputee was fit with a myoelectric prosthesis on the left side and body-powered device on the right. The pectoralis major and minor muscles on the left side were completely denervated during TMR surgery. However, when the user used strong biscapular abduction to operate the body-powered side, other muscles in the chest (perhaps the subclavius) produced unwanted EMG signals on the TMR side. Focused operation of the body-powered device reduced this problem but did not completely solve it.

6.3.5 Construction of the Final Socket

Carbon fiber braid is frequently used for its strength-to-weight characteristics in laminated sockets and frames. One problem with using carbon fiber is its conductivity. If remote electrodes are used in a flexible liner, motion can cause the remote electrode post to break through the outer plastic lamination, contact the carbon fibers, and cause electrical shorts. The same problem can occur if the remote electrode domes are placed in holes through the lamination. The additional electrodes necessary after TMR exacerbate this problem. Fiberglass reinforcement can provide similar strength while minimizing the chance of electrical problems in an already complicated system. In addition, because of the increased number of electrodes after TMR, we recommend that all inputs be clearly identified both at the connection to the socket and at the connection to the electronics/elbow. This precaution facilitates any future troubleshooting or repairs.

6.3.6 Component Selection

In general, the same considerations apply for selecting components after TMR as for a conventional fitting, including choosing to fit a hook, a hand, or both, and the possibility of using a powered wrist rotator. The user's desire to use certain terminal device components may influence the selection of the elbow, since not all terminal devices and elbow units are compatible. The goals and desires of the user should be balanced with hardware and software capabilities to select the best components for a specific individual. When it is known that an individual is considering a TMR procedure prior to initial fitting with a device, careful planning of componentry can help facilitate postsurgical fittings since not all devices are compatible with a TMR fitting.

Currently, the three main manufacturers of externally powered prosthetic elbows (Liberating Technologies Inc., Motion Control Inc., and Otto Bock HealthCare) all make systems that are compatible with prosthetic fitting after TMR. Discussing fitting plans and user requirements or preferences with the various manufacturers of prosthetic devices may be helpful. Issues to consider in the decision-making process include:

1. The ability to use remote electrodes. In addition to allowing easier electrode repositioning, using remote electrodes allows the two contacts to be angled in different directions, for example around a curve. This may allow better maintenance of electrode-skin contact in shoulder disarticulation fittings with excessive soft tissue movement.

2. The ability to visualize all of the signals during myotesting. It is useful to be able to see all signals at once when programming the device so that co-contraction can be identified and accommodated.

3. The flexibility of the software to adapt to control changes. Some manufacturers provide custom software that requires different input configurations, rather than a more adaptable "plug and play" type system.

6.3.7 Follow-up and Training

The standard of care for any medical intervention, including prosthetic fittings, should include appropriate follow-up. This is especially true with TMR fittings. Many factors, such as donning and doffing, basic operation, mode selection, hygiene, care, and maintenance, must be perfected by users so that they can take full advantage of the use of their prostheses. Proper follow-up care will help ensure that users continue to obtain the full functional benefits provided by TMR.

Electrode locations may shift slightly during transfer from a check socket to a final socket and frame construction. Because very specific electrode locations are needed to ensure proper control, it will often be necessary to fine-tune electrode gains and thresholds to optimize control in the final fitting.

After gaining experience with the device, users may wish to make changes to the control strategy. It may also be possible to change the control of a wrist rotator from an external input or an external switch to using EMG control of hand and wrist using either co-contraction or rate-sensitive mode selection. Also, signal strength and isolation may change as reinnervation continues to occur, and it may be desirable or necessary to move myoelectric sites.

Functional performance may be improved by changing electrode gains or thresholds. However, assuming appropriate electrode locations, improved performance often depends on users being able to generate sufficiently strong, separate signals. Thus, rather than making frequent adjustments to software in an attempt to optimize function, users should first practice exercises to strengthen and isolate EMG signals. (See Chapter 7 for recommended exercises.)

Working closely with an occupational therapist during fitting, training, and troubleshooting is ideal. There are times when only the prosthetist will be able to rectify a problem; for example, when electrode lift-off occurs. However, knowing which electrodes control which motor enables the occupational therapist to troubleshoot problems. The occupational therapist can also determine if the device operates differently or inadvertently in certain positions and provide this information to the prosthetist; for example, when the arm is overhead, the different torque of the socket about the limb may affect electrode-skin contact.

6.4 Conclusion

Although many standard fitting principles apply to individuals before and after TMR, there are unique issues and challenges that must be faced. The prosthetist must identify the optimal location of additional control sites and place additional electrodes into a well-fitting socket. The control achieved by individuals with TMR can exceed that of individuals with a conventional fitting. However, users must be willing to practice and strengthen the new control sites, and good clinical teams must work together with users to reach this level of success.

References

Malone, J. M., L. L. Fleming, J. Roberson, et al. 1984. Immediate, early, and late post-surgical management of upper-limb amputation. *J Rehabil Res Dev* 21 (1):33–41.

Miguelez, J.M., and M. D. Miguelez. 2003. The MicroFrame: the next generation of interface design for glenohumeral disarticulation and associated levels of limb deficiency. *J Prosthet Orthot* 15 (2):66–71.

Miguelez, J. M., M. D. Miguelez, and R. D. Alley. 2004. Amputations about the shoulder: prosthetic management. In *Atlas of Amputations and Limb Deficiencies*, 3rd ed., edited by D. Smith, J. W. Michael, and J. H. Bowker. Rosemont, IL: Amercan Academy of Orthpaedic Surgeons.

Pezzin, L. E., T. R. Dillingham, E. J. MacKenzie, P. Ephraim, and P. Rossbach. 2004. Use and satisfaction with prosthetic limb devices and related services. *Arch Phys Med Rehabil* 85 (5):723–729.

Sauter, W. F. 1992. Experience with electrically powered and myoelectric control systems in upper extremity prosthetics for children and adults. *Med Orthop Tech* 112:13–16.

7

Occupational Therapy for the Targeted Muscle Reinnervation Patient

Kathy A. Stubblefield and Todd A. Kuiken

CONTENTS

7.0 Introduction

Strategic occupational therapy of the targeted muscle reinnervation (TMR) patient before and after surgery is essential for optimizing the patient's eventual functional ability. Many tools and processes are available for the successful assessment and training of TMR patients. In addition, there are several issues that the therapist must consider throughout the course of treatment, many of which are unique to TMR patients and may be new to the therapist. This chapter is intended to serve as an evidence-based reference for therapists treating TMR patients before and after surgery. Here, we recommend preoperative assessment and education, a postoperative home program, follow-up visits prior to the TMR fitting, and a sequence for training the individual to use TMR control of standard-of-care prosthetic devices. Periodic assessment of function and progress using currently available outcome measures is also encouraged. The goal of TMR is to improve control of above-elbow prostheses. Therapists help patients realize this goal by ensuring mastery of prosthesis control through appropriate education and training.

7.1 Presurgical Program

7.1.1 Presurgical Evaluation

A thorough musculoskeletal and sensory exam is required for any patient prior to TMR surgery (see Chapter 5). Special attention should be given to the proximal shoulder on the injured side, as proximal kinetic chain injury is very common with traumatic arm amputation. A thorough evaluation of the patient's overall posture and core conditioning is also important, as postural musculature plays a significant role in the control of upper limb myoelectric prostheses. In addition, an assessment should be made of the patient's enthusiasm for using a prosthesis, compliance with past treatment, and understanding of the TMR process. If the patient has already been fit with a prosthesis, the patient's functional ability with the device should be thoroughly assessed before surgery to provide an indication of prosthesis control prior to TMR surgery and allow comparison to functional control after TMR. This baseline assessment cannot be performed after surgery, as prosthesis control may degrade before the final TMR fitting, due to denervation of original control muscles.

The consistent use of appropriate outcome measures is important for documenting changes in the patient's functional ability with a prosthesis before and after TMR. Many measures of upper limb performance are available to assess different aspects of prosthesis use by children (Wright 2006; Miller and Swanson 2009). However, few of these measures are specifically designed or validated to assess prosthesis use in adults (Wright 2006). No single outcome measure can fully assess prosthesis functionality (Metcalf et al. 2007; Lindner et al. 2010); each measure evaluates different aspects of prosthesis use, such as control, task performance, participation in activities, or patient satisfaction. Thus, combining a series of tests into a "toolkit" of appropriate measures within the framework of function, activity, and participation provides a more complete picture of a user's ability (Wright 2006). Box 7.1 lists a number of performance metrics used to quantify upper limb prosthesis use. Although many of these tests are validated with other patient populations, none of them has yet been validated for use with upper limb amputees.

7.1.2 Presurgical Education and Training

Prior to surgery, the therapist and patient should review techniques for maximum independence in basic self-care without a prosthesis in preparation for the immediate postsurgical period, during which the patient cannot wear a prosthesis. This review should include one-handed techniques and the use of adaptive equipment (see, e.g., Figure 7.1). However, it is important that the individual continue to use a prosthesis as much as possible before surgery and after surgical healing because of the significant potential for learned disuse (Malone et al. 1984; Atkins 2004; Taub et al. 2006). Early prosthetic fitting correlates with successful long-term integration of the prosthesis (Malone et al. 1984). Early training in prepositioning the terminal device (i.e., orientation of the prosthesis components in space in the position the natural hand would assume prior to beginning a task) and integrating the prosthesis into daily activities will carry over to the use of the TMR prosthesis and will promote functional independence.

Expectations for surgery, recovery, muscle reinnervation, prosthetic fitting, and prosthetic training should be thoroughly discussed with the patient and his or her family. While the physicians will provide much of this information, it is important that the patient hears the plan repeatedly and from different perspectives. Detailed information regarding surgical expectations and the general rehabilitation plan are described in Chapters 3 and 5, respectively. The therapist should understand the material well and be able to address basic questions. New medical issues should be referred to the treating physician. The therapist should explain the postoperative occupational therapy program to the patient in sufficient detail, based on the timing and frequency of planned follow-up visits, so that the patient is fully informed of the changes expected between appointments. The patient should be provided with both written and visual information, such as handouts or a video of required exercise programs, to take home.

While a linear progression is described here to explain the different aspects of the process (Figure 7.2), it is likely that timelines will vary depending on clinical variability between patients. Treatment phases may be longer or shorter, and may overlap, depending on the needs and progress of the individual.

BOX 7.1

Overview of quantitative and qualitative measurement tools appropriate for inclusion in a toolbox for assessment of prosthesis use in adults before and after TMR. Measures may be diagnosis-specific (developed for use with prosthesis users) or non-diagnosis-specific (developed to assess general upper limb function or manual dexterity in different subject populations). Qualitative measures are likewise either specific to prosthesis users or are general measures of quality of life.

Quantitative Outcome Measures	Qualitative Outcome Measures
Southampton Hand Assessment Procedure (SHAP): An objective test of unilateral hand function that can be used to evaluate functionality of passive, mechanical, or myoelectric hands without bias to type. Movement of abstract objects (some light, some heavy; classified into six prehension patterns) and activities of daily living are timed by the subject. Scores are compared to a normalized, able-bodied control score of 100. (Light et al. 2002)	**Orthotics and Prosthetics User Survey– Upper Extremity Functional Status (OPUS–UEFS)**: A 19-item, self-reported measure of an individual's ability to perform self-care and upper limb–based daily living tasks. (Burger et al. 2008)
Assessment of Capacity for Myoelectric Control (ACMC): An observational assessment tool that measures the subject's ability to control a myoelectric hand. The ability to control gripping, holding, releasing, and coordinating 30 items is scored on a 4-point capability scale. The prosthetic hand is used in an active assist or passive support role. The test can detect small differences in quality of control during performance of a bimanual task. (Hermansson et al. 2005; Lindner et al. 2009)	**Trinity Amputation and Prosthesis Experience Scales (TAPES):** A 54-item self-reported health-related quality-of-life questionnaire designed for adults with upper or lower limb amputation. It measures three aspects of wearing a prosthesis: psychosocial adaptation, activity restrictions, and satisfaction. (Gallagher and MacLauchlan 2000)
UNB Test of Prosthetics Function: An observational measure using dual 5-point rating scales for both function and spontaneous prosthesis use in children ages 2–13. It has also been adapted for use in children ages 14–21. (Sanderson and Scott 1985; Lake 1997)	**Canadian Occupation Performance Measure (COPM):** An individualized assessment in the areas of self-care, productivity, and leisure, using a semi-structured interview in a five-step process that measures subject-identified priorities in daily activities and generates two scores: one for performance and one for satisfaction with performance. (Law et al. 1990; Chan and Lee 1997)
	Disabilities of the Arm, Shoulder, and Hand Outcome Measure (DASH): An outcome measure developed for patients with general upper-extremity musculoskeletal conditions. This 30-item, self-reported questionnaire aims to both describe extent of ability (physical function and symptoms) and changes in these parameters over time. (Hudak et al. 1996; Beaton et al. 2001; Beaton et al. 2005)

(continued)

Quantitative Outcome Measures	Qualitative Outcome Measures
Box and Block Test: A timed test devised to evaluate gross manual dexterity with normative data for individuals age 6 and older. Small (2.5-cm) blocks are moved one at a time from one side of a box to the other over a wooden partition. The score is the number of blocks transferred in 1 min. (Mathiowetz et al. 1985) **Jebsen-Taylor Hand Function Test**: A seven-part standardized timed test of manual dexterity that evaluates hand function using common items such as cans, paper clips, and coins. (Jebsen et al. 1969)	

FIGURE 7.1
Examples of adaptive devices that enable one-handed activities. (a) Button hook. (b) Rocker knife.

7.2 Postsurgical Program

7.2.1 Interim Prosthetic Issues and Activities

Most TMR patients will use a two-site myoelectric prosthesis prior to surgery and can resume use of their prosthesis after surgery when there is adequate healing of the incision and the postoperative edema has abated. TMR surgery does not affect the control of body-powered prostheses; however, fat removal or soft tissue revision may alter the size and shape of the residual limb. This may result in the need for socket modification or replacement. Myoelectric or hybrid prosthesis users with transhumeral amputations should be able

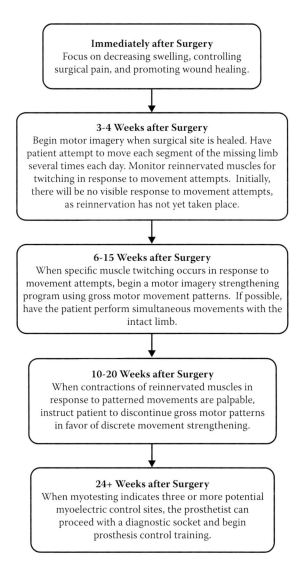

FIGURE 7.2
General timeline for occupational therapy after TMR.

to continue using a two-site myoelectric or hybrid prosthesis after surgery, with electrodes situated over the undisturbed heads of the biceps and tri-ceps muscles for elbow control. However, as the optimal electromyographic (EMG) recording locations may have changed, the electrodes may need to be repositioned. Shoulder disarticulation TMR surgery will denervate and thus paralyze most of the useable chest muscles. After TMR, these patients will likely need a control system that does not use EMG signals. Shoulder straps with built-in switches, shoulder rocker switches, and force-sensitive resistors can be used along with body-powered components. Unilateral shoulder disarticulation patients may choose not to wear a prosthesis during reinnervation because of poor function; this is a reasonable and acceptable choice. However, bilateral shoulder disarticulation patients will need an interim prosthesis with a modified control system.

If the patient continues to use a myoelectric prosthesis after surgery, prosthetic control should be modified so that the patient no longer uses unrelated muscles to control functions that will be intuitively controlled after reinnervation. For example, if biceps contraction was originally used to both flex the elbow and close the hand, body-powered control of hand closure may be substituted in the interim prosthesis to allow the patient to practice using the biceps only for elbow flexion. Use of a body-powered terminal device during reinnervation will reduce potential confusion later when training the patient to use a new and intuitive control site to close the hand. However, it may be reasonable and appropriate to use the patient's previous myoelectric control method if too many changes are frustrating for the patient.

The therapist and prosthetist should not assume that adequate prosthetic control after surgery will be automatic. Any alteration of the socket, control methods, or electrode locations will necessitate training to ensure that the patient can use the prosthesis as intended. Troubleshooting and training sessions may be required to ensure satisfactory control of function and comfort of the interim prosthesis. The therapist and prosthetist are responsible for ensuring that any changes made to the control of the interim prosthesis are understood by the patient, that all components are working, and that the socket fit is comfortable, so that the patient is able to continue wearing a prosthesis during reinnervation.

7.2.2 EMG Signal Strengthening and Patient Education

Success with prosthesis control after TMR is in part dependent on following a home program to obtain strong, independent, and isolated EMG signals from reinnervated muscles. The program involves a progression of exercises that will facilitate natural and seamless control of the device. After a 2- to 3-week period to allow the surgical incision to heal and the soft tissues to consolidate, patients should start a program to "exercise" their amputated arm. Patients are instructed to try to move all of their missing joints: the elbow, wrist, hand, and digits. This *mental practice*—the repetitive motor imagery of a given movement—should be done several times a day. Mental practice generates changes in brain activation (Lafleur et al. 2002; Jackson et al. 2003), improves motor performance (Yue and Cole 1992), and can increase muscular force (Cornwall et al. 1991) without direct muscle activation, presumably through central mechanisms (Yue and Cole 1992; Ranganathan et al. 2004). These movement attempts may promote both central and peripheral nerve pathways to enhance reinnervation and will prime the patient to recognize the first signs of reinnervation: small muscle twitches in response

TABLE 7.1

Native Nerve Actions of the Upper Limb

Musculocutaneous Nerve	Radial Nerve	Median Nerve	Ulnar Nerve
Elbow flexion	Elbow extension	Pronation	Ulnar deviation
	Forearm supination	Wrist flexion	Ulnar finger flexion
	Wrist extension	MP[a] and PIP[b] flexion	Finger abduction
	Radial deviation	Thumb flexion	
	Finger extension		
	Thumb extension		

[a] Metacarpophalangeal joint.
[b] Proximal interphalangeal joint.

to attempted movement, which occur approximately 10 to 15 weeks after surgery. Tingling sensations and muscle twitching, independent of movement attempts, may occur during early reinnervation.

Strengthening of relevant muscles for myoelectric prosthesis control is required following any amputation, but different exercises are required after TMR because of the large amount of motor control information transferred to the target muscles. Each large transferred motor nerve contains motor control content for a variety of arm and hand functions (Table 7.1) and reinnervates a relatively small area of muscle. The potential movements controlled by each nerve will be represented to varying degrees in the target muscle. However, reinnervation may be disproportionate such that the EMG signals for some movements may be large and other movements may produce little or no EMG signal. Until reinnervation is complete, neither the surgeon nor the therapist can be certain of which limb movements will be represented or which may be best used for prosthesis control. As a result, the therapist must thoroughly understand peripheral nerve function and know, based on the individual surgical procedure, which nerves are anticipated to reinnervate which muscle regions.

7.2.2.1 Gross Motor Patterns

When the first noticeable signs of reinnervation—muscle twitches under voluntary control—are detected, the individual should begin visualizing and attempting movement patterns corresponding to the peripheral nerve distribution. This process is known as *motor imagery*—an active mental representation of movement involving sensation and perception without motor output or actual body movement (Jackson et al. 2001; Dickstein and Deutsch 2007). Motor imagery activates areas in the brain similar to those involved in actual movement (Leonardo et al. 1995; Stephan et al. 1995; Roth et al. 1996; Ehrsson et al. 2003) and results in increases in heart rate, blood pressure, and respiration rate in correlation with the extent of imagined effort.

FIGURE 7.3
Examples of gross motor patterns: flexion (median and ulnar nerves, top) and extension (radial and ulnar nerves, bottom).

The goal of motor imagery and mental practice at this point is to facilitate and reinforce both central control mechanisms and peripheral reinnervation processes corresponding to the hand and arm movements controlled by transferred nerves. Mass movement patterns involving flexion and extension (Figure 7.3) and incorporation of the motor repertoire of each transferred peripheral nerve are used. Performing mirrored movements with the intact limb provides visual and kinesthetic reinforcement.

For shoulder disarticulation TMR patients, elbow flexion should be exercised separately from forearm and hand patterns. For transhumeral TMR patients, the elbow should remain relaxed during patterned exercises. Experience has shown that the coupling of elbow and hand actions makes it difficult to achieve four isolated control sites at the time of the TMR fitting.

Because resistance cannot be applied to a missing limb, strength training is achieved by progressively increasing the force and duration of muscle contractions, the number of repetitions in each set, and the frequency of exercise. Exercise sets should be repeated at least four times per day to rein-force motor pathways and increase muscle strength, as the reinnervated muscle will be atrophied from temporary paralysis. Early in training, the muscles will fatigue quickly and will need time to recover. Patients should be taught to recognize fatigue and take frequent rest breaks between con-tractions. The therapist can apply resistance to the residual humerus or scap-ula to strengthen the supporting structures of the shoulder as necessary. The patient should also be given a core-strengthening program to ensure good posture and help support prosthesis use.

7.2.2.2 Discrete Actions

When consistent target muscle activation is palpable during gross movement exercises (usually 3 to 5 months after surgery), the patient should discontinue the gross movement patterns and begin to attempt discrete actions, such as closing the hand independently from wrist movement (median nerve). The objective here is to develop control actions that are strong, independent (i.e., separate for each prosthesis movement), and isolated (i.e., without coacti-vation of other muscles) in preparation for fitting the new TMR-controlled prosthesis. Actions are attempted in isolation, keeping joints distal and proximal to the axis of movement relaxed (e.g., keeping the elbow and fin-gers relaxed while attempting wrist extension). In the absence of visual or proprioceptive feedback, attempting to perform these isolated actions can be difficult. It can be helpful to encourage the individual to perform the actions bilaterally. A mirror (or a mirror box, as used to treat phantom limb pain [Ramachandran and Rogers-Ramachandran 1996]), which allows the patient to watch the intact limb perform the intended action while attempting the same action with the missing limb, may be useful. This continued strength-ening of the target muscles helps the patient develop the endurance needed for training with and using the prosthesis. As muscles become stronger, the patient should begin graded contraction exercises to allow for eventual use of proportional control.

The following cues may be helpful in enabling the patient to attempt nerve-specific movements:

Proximal radial nerve transfer (elbow extension): Patients can visualize pushing down on armrests while rising from a chair or perform-ing push-ups against a wall. Posterior deltoid activity, as involved in shoulder extension, should be discouraged.

Musculocutaneous nerve transfer (elbow flexion): Patients can visual-ize reaching for their chin or lifting a bag with the handle over their forearm while keeping their hand relaxed and their forearm neutral.

Distal radial or ulnar nerve transfer: Patients can visualize spreading their fingers (hand open), bringing the wrist into extension while

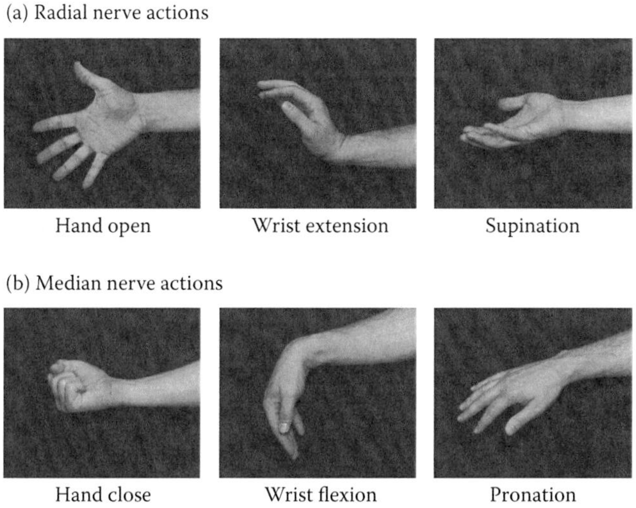

(a) Radial nerve actions

| Hand open | Wrist extension | Supination |

(b) Median nerve actions

| Hand close | Wrist flexion | Pronation |

FIGURE 7.4
Discrete movements, showing actions controlled by the radial nerve (a) and the median nerve (b), practiced as part of a typical home program to develop isolated EMG control sites for prosthesis control. (Adapted from Simon et al., Patient training for functional use of pattern recognition–controlled prostheses, *J Prosthet Orthot* 24 (2):56–64, 2012. With permission.)

keeping the fingers relaxed (wrist extension), and rotating the forearm toward "palm up" (supination) (Figure 7.4a).

Median or ulnar nerve transfer: Patients can visualize making a fist (hand close), bending their wrist into flexion while keeping their fingers relaxed and their forearm neutral with regard to rotation (wrist flexion), or turning their palm down by leading with thumb flexion/abduction (pronation) (Figure 7.4b).

7.2.2.3 *Muscle Relaxation*

The patient must learn to incorporate muscle relaxation into the exercise program. This can also be achieved using motor imagery. Muscle relaxation is important for avoiding overexertion and fatigue, which frequently result in inadvertent co-contraction, muscle cramping or spasm, and loss of useful EMG signals. In addition, muscle relaxation is an important component of myoelectric prosthesis control, as it allows the EMG signal to fall below a certain threshold and is used as an "off" signal. The therapist should encourage individuals to palpate their reinnervated area while attempting exercises to determine whether muscles are contracted (hard) or relaxed (soft). A portable myotesting device can also be used to give patients feedback as to how well they are relaxing.

7.2.2.4 Phantom Limb Sensation

Most amputees experience phantom limb sensation or awareness, which may include pain or discomfort. Some patients perceive movement of the phantom limb in response to attempted movements of the missing limb; in this case, phantom limb sensation may be helpful for prosthesis control. Frequently, however, the phantom limb is immobile (feels paralyzed). In this case, phantom sensation can confuse patients and cause difficulty when they attempt exercises. Patients with this experience must learn to distinguish between attempts to move the amputated limb and movement (or nonmovement) of the phantom limb. They must understand that phantom limb position is unrelated to prosthesis operation. It is important to teach patients that the brain sends signals through the transferred nerves in response to attempted movements regardless of phantom limb position or mobility. As the transferred nerve fibers grow into the target muscles, these signals cause motor units to fire, resulting in contractions of individual target muscle fibers that are at first imperceptible. As a result, the therapist should instruct patients to perform the exercises even if they feel as if the phantom limb is not moving, because the resulting nerve signals will cause the reinnervated muscles to contract and strengthen. Performing exercises simultaneously with the intact limb encourages appropriate attempted movements from the patient and decreases distraction from the phantom limb.

Phantom limb pain such as cramping, clenching, or feelings of limb paralysis may interfere with attempts to exercise reinnervated muscles. Any phantom limb pain should be discussed with the patient's physician. Treatments include an array of neuropathic pain medications, limb massage or tapping, stress relaxation techniques, biofeedback, and mirror therapy (Ramachandran and Rogers-Ramachandran 1996; Chan et al. 2007).

7.2.3 Diagnostic Fitting and Four-Site Training

7.2.3.1 Diagnostic Socket Development

Approximately 6 months after surgery, when target muscles are sufficiently reinnervated, the prosthetist will identify EMG recording sites for the diagnostic socket. It is important that the team be patient and not fit the TMR prosthesis too soon, as optimal control sites are likely to change as reinnervation progresses. It can be frustrating for the patient and clinical team if control sites are selected too early: system performance may deteriorate significantly as reinnervation progresses and signals strengthen. Control sites may still change after 6 months, requiring reevaluation of control sites or muscle actions for control; however, these changes will not be as fast or as profound as in the initial reinnervation stage.

Knowledge of the location of each brachial plexus nerve transfer and the functions of each nerve is essential for developing the optimal control strategy. Patients with transhumeral amputations can expect independent, intuitive control of elbow flexion and extension from the long heads of the biceps and triceps innervated by the intact native musculocutaneous and proximal radial nerves, respectively. The lateral head of the triceps, reinnervated by the distal radial nerve, and the short head of the biceps, reinnervated by the median nerve, can be used for intuitive control of hand open and hand close, respectively. For patients with long transhumeral amputations, the ulnar nerve transfer to the brachialis muscle can potentially control hand open, hand close, or wrist rotation (see Table 7.1). Patients with shoulder disarticulations can expect intuitive control of elbow flexion/extension and hand open/close from four myoelectric control sites on reinnervated target muscles. Other muscles (such as remnant deltoid) or body-powered or electromechanical switching (e.g., pressure sensors or pull switches) can be used to control other joint movements, such as wrist rotation or a shoulder lock.

Thorough EMG testing of attempted movements is the next step in configuring the diagnostic TMR prosthesis. Beginning with a single degree of freedom, such as elbow flexion/extension, the prosthetist will observe muscle EMG activity and adjust electrode gain settings for optimal signal strength and balance. Initial EMG signals from target muscles will be weak, with limited endurance. Patients should be asked to describe in detail each attempted movement by demonstrating with their intact limb. Other methods of attempting the same movement can then be explored. The goal of this process is to identify a physiologically relevant movement that provides the strongest, most independent EMG signals. However, the most intuitive attempted movement may not generate sufficient EMG signals; for example, finger extension is the most intuitive control movement for hand open, but a less intuitive movement, such as wrist or thumb extension, may provide a stronger EMG signal. In determining the optimal intended movement for each prosthesis function, the signal separation between attempted movements for different functions must also be taken into account. Sites that generate the strongest EMG signals may be too close to other control sites. For example, for hand open, wrist extension may generate a stronger EMG signal than finger extension, but finger extension may provide better signal separation from elbow extension.

A diagnostic socket can be prepared when the therapist and prosthetist are satisfied that they have identified adequate, distinct EMG signals for control of each prosthetic function. While the socket is being fabricated, training should focus on the specific attempted actions that generate the required signals so that each action becomes easy and natural, facilitating future control of the prosthesis. Therapists can use feedback from EMG visualization tools, such as a myotester, or remote control of a prosthesis to optimize independence of attempted movements and help develop differentiated signals. These feedback methods will also train the patient to recognize the relaxed state, which is essential for successful control. The diagnostic socket should be used for several weeks during initial training to ensure optimal fit and function.

7.2.3.2 Initial TMR Prosthetic Training

In many ways, training an amputee to use a TMR prosthesis is easier than training him or her to use a conventional prosthesis. With TMR, the patient uses natural neural pathways to operate the elbow and hand—no switching is needed between these two fundamental movements. Training can often progress quickly if (1) the prosthesis has been fit well, (2) there are four distinct EMG control signals, and (3) the patient has conditioned both the reinnervated muscles and core strength.

An objective of the initial training process is to verify electrode locations and to allow for alternative site exploration if necessary. The addition of a socket and the weight of the prosthesis will likely affect the characteristics of the EMG signals, necessitating adjustments. The initial goal is for the patient to demonstrate clearly differentiated flexion and extension signals and be able to relax between contractions. Additionally, the patient will develop tolerance for the weight of the prosthesis and become accustomed to the movement of the components, with respect to both speed and sound. The patient must also learn to recognize fatigue in the context of wearing the new myoelectric prosthesis.

Once control of each degree of freedom is mastered, the therapist will help the patient achieve separate control of two degrees of freedom (elbow flexion/extension and hand open/close). The goal here is to achieve isolated, independent movement of each degree of freedom. Although TMR allows the patient to have simultaneous control of both degrees of freedom, it is important that the patient be able to perform each in isolation so that accidental activation of one does not occur during the intended performance of the other. Inability to perform isolated movements may be due to co-contraction, suboptimal electrode placement, or the size of target muscle EMG signals compared to signals from adjacent muscles (i.e., muscle cross talk). It can be difficult initially to determine the cause of the aberrant movement. Frequent adjustments to electrode gains, thresholds, electrode locations, and software are often necessary to obtain adequate signal separation.

Early prosthesis training with performance drills of each action will reinforce the physiologically appropriate control movements. Practice strengthens target muscles and EMG signals, minimizes co-contractions, and encourages separation of control signals for different prosthetic functions. Practicing elbow movements with the hand partially open, and hand functions with the elbow and wrist in a neutral position, allows unwanted movements to be detected and corrected. It may be helpful for the patient to mirror movement intentions using the intact limb. To further evaluate control, it is helpful if the patient verbally indicates the movement that is intended before each attempt. Complexity can be added by having the patient change the position of the shoulder or move about in the environment while controlling the prosthesis.

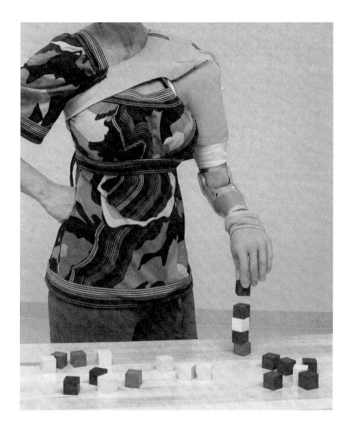

FIGURE 7.5
A transhumeral subject performs a block-stacking task.

The next goal is to train the patient to transition quickly and easily from one prosthetic function to another. The patient and therapist should first focus on increasing the speed of movement drills. The patient can then begin using the prosthesis to perform tasks. As always, the patient should start with moving slowly through simple tasks, such as stacking blocks or cans (Figure 7.5), then progress to more challenging tasks and functional activities of daily living (ADLs).

7.2.3.3 Additional Prosthetic Components and Alternative Control Options

Once the patient demonstrates consistent, independent control using four myoelectric sites, additional prosthetic features can be added. The most common additions are a powered wrist rotator and/or a shoulder lock. There are a variety of switches, buttons, and EMG strategies for operating these components. Exercises should be implemented (as previously discussed) to allow the patient to master control of these new features without compromising control of previously mastered functions. The patient must be trained to activate unfamiliar controls, (e.g., a linear transducer pull-switch to provide wrist rotation), without contracting other muscles involved in prosthesis control. In this case, the patient must be able to use the switch without causing unwanted hand or elbow activity. Likewise, when using a powered chin switch to lock and unlock the shoulder, it is necessary to be able to depress the switch without activating the chest muscles that control the distal prosthesis components.

7.2.3.4 One-Handed, Bimanual, and Advanced Functional Activities

When the patient has mastered independent control of all prosthetic functions, then training in one-handed, bimanual, and other advanced activities may begin. Functional training is gradually introduced by having the patient practice grasping and releasing common objects. Patients can practice holding objects while moving proximal joints of the prosthesis, moving the contralateral limb, and moving around in their environment. Patients can also practice passing objects from hand to hand and holding objects with the prosthesis while manipulating them with the contralateral limb.

Training in one-handed functional activities encourages the patient to explore the full functionality of the prosthesis and possibly to discover additional abilities using their TMR control that they might not otherwise notice because of preexisting habits in performing routine tasks. Tasks such as loading or unloading a dishwasher, sorting mail, or unloading a dryer all have elements of repetition, involve all three arm functions, and, when performed one handed, require some prepositioning. Inadequate prepositioning is a frequent cause of inability to complete tasks easily or well. Experience with repetitive, patterned, functional tasks further reinforces the appropriate motor commands. These tasks also provide an opportunity to practice use of the prosthesis in daily activities.

Tasks requiring bimanual coordination, such as zipping a jacket, folding clothes, and removing money from a wallet, increase the complexity and prepositioning demands (Figures 7.6 and 7.7). Instruction and possible reeducation in body mechanics may be required to help the patient gain maximum benefit from the prosthesis. Timing the tasks may be used to motivate the patient to focus on and improve performance.

Once the patient is competent in basic operation of the final limb system, more challenging ADLs, vocational tasks, and avocational activities can be attempted. This is also the time to address any additional unique functional goals and tasks that the patient considers important. Examples we have frequently encountered include fishing, shooting/hunting, and using a variety of tools. Sometimes adaptive equipment is needed. Often it is primarily a matter of helping a patient work through the steps of the activity and solve problems.

FIGURE 7.6
A transhumeral subject uses a TMR prosthesis to assist in zipping a jacket. (Adapted from Simon et al., Patient training for functional use of pattern recognition–controlled prostheses, *J Prosthet Orthot* 24 (2):56–64, 2012. With permission.)

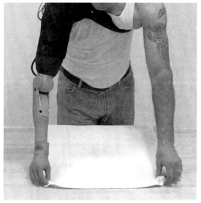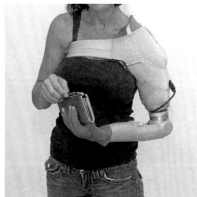

FIGURE 7.7
Two transhumeral subjects use their TMR prostheses to perform bimanual activities. (*Left:* Adapted from Simon et al., Patient training for functional use of pattern recognition–controlled prostheses, *J Prosthet Orthot* 24 (2):56–64, 2012. With permission.)

7.2.3.5 *Taking the Prosthesis Home*

The patient is sent home with the diagnostic socket and prosthesis when the team agrees that technical problems have been overcome, control is adequate, and the patient can troubleshoot and resolve most issues that may arise. If possible, the time between successive appointments should be kept short, allowing just a few days for practice using the device at home. It may be helpful to have the patient generate a list of activities in which he or she will try to use the prosthesis. These tasks can be practiced in the clinic, and the therapist can offer suggestions for postural cues and prepositioning to prevent development of bad habits. It is useful to give the patient a troubleshooting guide, (e.g., what to do if there is unintended activity in the prosthesis), and to follow up at the next clinic visit to help resolve any complaints. When embedded in a complex task, simple bimanual activities will present a challenge for the patient and issues may arise that went unnoticed in the structured clinic environment.

7.2.3.6 *Simultaneous Control*

When patients have good control of each function individually, they can start to perform combined actions: simultaneous control of two different functions. Unlike conventional myoelectric control, TMR allows simultaneous control of the hand and elbow. Wrist rotation can also be simultaneously controlled with some type of shoulder switch. Thus the therapist should provide cues encouraging the patient to perform simultaneous, combined movements, such as opening the hand while reaching for an object. Instruction as to proper prepositioning of the shoulder and elbow joint and possible relearning of body mechanics may be required to allow the patient to gain maximum benefit from the prosthesis. Timing the tasks may be used as a motivator to focus on and improve performance. Speed does not signify success but can be used during training to develop simultaneous or seamless-sequential control. This kind of practice will demonstrate to the patient the full potential of TMR control.

It is important that the patient retains the ability to independently control each separate function in addition to learning to perform them as part of movement combinations: opening the hand during elbow extension is useful for reaching to pick up a cup but not for reaching to place a full cup of water on a table. In the latter case, hand open must be separated, not patterned, with elbow extension. If combined movement training results in unwanted movements of other degrees of freedom, the process should be slowed down and simplified to allow the patient to reestablish independent control.

7.3 Conclusion

The success of the TMR procedure is dependent on the skill, creativity, knowledge, collaboration and hard work of the prosthetist, therapist, patient, and family. The course is fairly lengthy, and an important part of the therapeutic process must be to sustain enthusiasm and motivation. This is best achieved through working with the patient frequently, having shared objectives, and good communication. The hard work is well worth it, as it can result in significant improvement in prosthesis control and function and, therefore, the patient's overall quality of life.

References

Atkins, D. J. 2004. Functional skills training with body-powered and externally powered prostheses. In *Functional Restoration of Adults and Children with Upper Extremity Amputation*, edited by R. H. Meier and D. J. Atkins. New York, NY: Demos Medical Publishing, Inc.

Beaton, D. E., J. N. Katz, A. H. Fossel, J. G. Wright, V. Tarasuk, and C. Bombardier. 2001. Measuring the whole or the parts? Validity, reliability, and responsiveness of the Disabilities of the Arm, Shoulder and Hand outcome measure in different regions of the upper extremity. *J Hand Ther* 14(2):128–46.

Beaton, D. E., J. G. Wright, and J. N. Katz. 2005. Development of the QuickDASH: comparison of three item-reduction approaches. *J Bone Joint Surg Am* 87(5):1038–46.

Burger, H., F. Franchignoni, A. W. Heinemann, S. Kotnik, and A. Giordano. 2008. Validation of the orthotics and prosthetics user survey upper extremity functional status module in people with unilateral upper limb amputation. *J Rehabil Med* 40(5):393–9.

Chan, B. L., R. Witt, A. P. Charrow, et al. 2007. Mirror therapy for phantom limb pain. *N Engl J Med* 357 (21):2206–2207.

Chan, C. C. H., and T. M. C. Lee. 1997. Validity of the Canadian Occupational Performance Measure. *Occup Ther Int* 4(3):231–49.

Cornwall, M. W., M. P. Bruscato, and S. Barry. 1991. Effect of mental practice on isometric muscular strength. *J Orthop Sports Phys Ther* 13 (5):231–234.

Dickstein, R., and J. E. Deutsch. 2007. Motor imagery in physical therapist practice. *Phys Ther* 87 (7):942–953.

Ehrsson, H. H., S. Geyer, and E. Naito. 2003. Imagery of voluntary movement of fingers, toes, and tongue activates corresponding body-part-specific motor representations. *J Neurophysiol* 90 (5):3304–3316.

Gallagher, P., and M. MacLauchlan. 2000. Development and psychometric evaluation of the Trinity Amputation and Prosthesis Experience Scales (TAPES). *Rehabil Psychol* 45(2):130–54.

Hermansson, L.M., A. G. Fisher, B. Bernspang, and A.C. Eliasson. 2005. Assessment of capacity for myoelectric control: a new Rasch-built measure of prosthetic hand control. *J Rehabil Med* 37(3):166–71.

Hudak, P. L., P. C. Amadio, and C. Bombardier. 1996. Development of an upper extremity outcome measure: the DASH (disabilities of the arm, shoulder and hand) [corrected]. The Upper Extremity Collaborative Group (UECG). *Am J Ind Med* 29(6):602–8.

Jackson, P. L., M. F. Lafleur, F. Malouin, C. Richards, and J. Doyon. 2001. Potential role of mental practice using motor imagery in neurologic rehabilitation. *Arch Phys Med Rehabil* 82 (8):1133–1141.

Jackson, P. L., M. F. Lafleur, F. Malouin, C. L. Richards, and J. Doyon. 2003. Functional cerebral reorganization following motor sequence learning through mental practice with motor imagery. *Neuroimage* 20 (2):1171–1180.

Lafleur, M. F., P. L. Jackson, F. Malouin, C. L. Richards, A. C. Evans, and J. Doyon. 2002. Motor learning produces parallel dynamic functional changes during the execution and imagination of sequential foot movements. *Neuroimage* 16 (1):142–157.

Lake, C. 1997. Effects of prosthetic training on upper-extremity prosthesis use. *J Prosthet Orthot* 9(1):3–9.

Law, M., S. Baptiste, M. McColl, A. Opzoomer, H. Polatajko, and N. Pollock. 1990. The Canadian occupational performance measure: an outcome measure for occupational therapy. *Can J Occup Ther* 57(2):82–7.

Leonardo, M., J. Fieldman, N. Sadato, et al. 1995. A functional magnetic resonance imaging study of cortical regions associated with motor task execution and motor ideation in humans. *Hum Brain Mapp* 3 (2):83–92.

Light, C. M., P. H. Chappell, and P. J. Kyberd. 2002. Establishing a standardized clinical assessment tool of pathologic and prosthetic hand function: normative data, reliability, and validity. *Arch Phys Med Rehabil* 83(6):776–83.

Lindner, H. Y., J. M. Linacre, and L. M. Norling Hermansson. 2009. Assessment of capacity for myoelectric control: evaluation of construct and rating scale. *J Rehabil Med* 41(6):467–74.

Lindner, H. Y. N., B. S. Natterlund, and L. M. N. Hermansson. 2010. Upper limb prosthetic outcome measures: review and content comparison based on International Classification of Functioning, Disability and Health. *Prosthet Orthot Int* 34 (2):109–128.

Malone, J. M., L. L. Fleming, J. Roberson, et al. 1984. Immediate, early, and late postsurgical management of upper-limb amputation. *J Rehabil Res Dev* 21 (1):33–41.

Metcalf, C., J. Adams, J. Burridge, V. Yule, and P. Chappell. 2007. A review of clinical upper limb assessments within the framework of the WHO ICF. *Musculoskeletal Care* 5(3):160–173.

Miller, L. A., and S. Swanson. 2009. Summary and Recommendations of the Academy's State of the Science Conference on Upper Limb Prosthetic Outcome Measures. *J Prosthet Orthot* 21:83–89.

Ramachandran, V. S., and D. Rogers-Ramachandran. 1996. Synaesthesia in phantom limbs induced with mirrors. *Proc R Soc Lond B Biol Sci* 263 (1369):377–386.

Ranganathan, V. K., V. Siemionow, J. Z. Liu, V. Sahgal, and G. H. Yue. 2004. From mental power to muscle power—gaining strength by using the mind. *Neuropsychologia* 42 (7):944–956.

Roth, M., J. Decety, M. Raybaudi, et al. 1996. Possible involvement of primary motor cortex in mentally simulated movement: a functional magnetic resonance imaging study. *Neuroreport* 7 (7):1280–1284.

Sanderson, E. R. and R. N. Scott. 1985. UNB test of prosthetics function, a test for unilateral upper extremity amputees. [test manual]. Bio-Engineering Institute, University of New Brunswick, Fredericton, New Brunswick, Canada.

Simon, A. M., B. A. Lock, and K. A. Stubblefield. 2012. Patient training for functional use of pattern recognition-controlled prostheses. *J Prosthet Orthot* 24 (2):56–64.

Stephan, K. M., G. R. Fink, R. E. Passingham, et al. 1995. Functional anatomy of the mental representation of upper extremity movements in healthy subjects. *J Neurophysiol* 73 (1):373–386.

Taub, E., G. Uswatte, V. W. Mark, and D. M. M. Morris. 2006. The learned nonuse phenomenon: implications for rehabilitation. *Eura Medicophys* 42 (3):241–256.

Wright, V. F. 2006. Measurement of functional outcome with individuals who use upper extremity prosthetic devices: current and future directions. *J Prosthet Orthot* 18 (2):46–56.

Yue, G., and K. J. Cole. 1992. Strength increases from the motor program: comparison of training with maximal voluntary and imagined muscle contractions. *J Neurophysiol* 67 (5):1114–1123.

8

Targeted Sensory Reinnervation

Paul D. Marasco

CONTENTS

8.0 Introduction

Targeted muscle reinnervation (TMR) was developed to provide intuitive myoelectric control sites for powered prosthesis users but additionally, and unexpectedly, resulted in the potential to provide sensory feedback. Several months after the first TMR surgery, the patient reported that touching the skin over the reinnervated muscles felt as if the missing hand or arm were being touched. It appeared that in addition to the reinnervation of muscles by motor fibers from the transferred mixed nerves, the overlying skin had been reinnervated by sensory fibers (Kuiken et al. 2004). We call this redirected sensation *transfer sensation*. Because nerve transfer can result in both motor and sensory reinnervation, we will use the term *targeted reinnervation* instead of TMR for the purposes of this chapter. With this exciting discovery came the potential for a new way to provide intuitive sensory feedback to prosthesis users. It also raised many questions about the qualities of transfer sensation, its capacity to provide useful feedback, and its underlying neural mechanisms. In this chapter, we review the experimental characterization of transfer sensation, the neural mechanisms involved, and potential applications of this interesting and exciting phenomenon.

8.1 Sensory Perception and Feedback in Prostheses

We rely heavily on our senses to perceive and interact with our environment. Our visual and somatosensory systems factor heavily into learning and perfecting tasks that involve moving and manipulating objects. The loss of an arm entails not only the loss of the bone, muscles, and nerves necessary for movement, but also of the various receptors and pathways of the somatosensory system used to relay information about that movement and the environment back to the brain. With the loss of this information, prosthesis users must rely exclusively on visual feedback to monitor the positioning and performance of their devices. The ramifications of loss of sensory perception for motor control are numerous, ranging from difficulty in performing fine motor tasks (due to the loss of touch perception) (Jenmalm and Johansson 1997; Zhang et al. 2011) to complete inability to control limb posture (as with systemic loss of touch and proprioception) (Gallagher and Cole 1995; Sainburg et al. 1995).

8.1.1 The Somatosensory System

The somatosensory system is distributed throughout the body and comprises the senses of cutaneous touch, proprioception (limb movement and position), nociception (pain), and thermoception (temperature). This information is encoded by various sensory receptors located in the skin, muscles, joints, and soft tissues. Four main types of cutaneous receptors appear to contribute most prominently to our perception of cutaneous touch: Merkel neurite complexes, Pacinian corpuscles, Meissner corpuscles, and slowly adapting type II (SA II) end organs. The inputs from these receptors are combined and integrated at multiple levels within the somatosensory system (Mountcastle 2005; Gescheider et al. 2009). Temperature and pain sensation are also integrated with tactile touch to provide detailed interpretations of the world that we experience, such as feeling that objects are wet or oily or recognizing touch as uncomfortable or pleasant. The interaction of multiple input channels to create the perception of touch is a particularly interesting aspect of the somatosensory system (Mountcastle 2005; Gescheider et al. 2009). However, this innate complexity also presents a challenge when working to restore a physiologically relevant and functional sense of touch to a prosthesis user.

To build our sense of touch, external inputs are detected by the distal axon terminal of a dorsal root ganglion neuron, which has its cell body located at the spinal cord. The distal axon terminals of the large-diameter myelinated fibers that carry signals for touch and proprioception are encased in the specialized cutaneus receptor structures mentioned above. In contrast, the small-diameter nonmyelinated and thinly myelinated neurons that detect pain and temperature have unencapsulated, free endings. Cutaneous touch and proprioceptive fibers ascend to the somatosensory cortex through connections in the dorsal column nuclei and thalamus. Pain and temperature fibers ascend to the brain through connections in the dorsal columns and thalamus (Mountcastle 2005). The somatosensory cortex is topographically arranged to form a representative map of the areas of the body where the receptors are located (somatotopic organization). The distribution of cortical area is not proportional to body surface area. Instead, greater cortical surface area is dedicated to more behaviorally relevant body parts, such as the fingertips and tongue.

8.1.2 Importance of Sensory Feedback for Prosthesis Control

Tactile feedback plays a critical role in our ability to understand the physical nature of objects and to manipulate them effectively (Ernst et al. 2000; Ernst and Banks 2002). In unimpaired individuals, the combination of visual and tactile feedback increases performance in object manipulation tasks (Huang et al. 2007). Sensory feedback also plays a role in the ability to learn new motor tasks and to modulate and refine active movement (Ernst and Banks 2002; Sober and Sabes 2005). In addition to their role in object manipulation, both tactile and proprioceptive feedback are fundamental to establishing our sense of body ownership and our conscious awareness of our actions (Fourneret et al. 2002; Armel and Ramachandran 2003; Ehrsson et al. 2004).

Because a prosthetic limb is an insensate tool, a user must rely almost exclusively on vision to monitor its operation. This results in slower, clumsier movements and a higher cognitive load—the user must continuously watch what the device is doing. There is a strong desire on the part of upper limb amputees to have prosthetic devices that provide sensory feedback (Atkins et al. 1996) and minimize the need for visual attention (Atkins et al. 1996; Biddiss et al. 2007).

In previous efforts to provide sensory feedback from prosthetic devices, researchers have utilized sensory substitution to transmit feedback signals from the device to alternative body surfaces (Beeker et al. 1967; Prior and Lyman 1975; Scott et al. 1980; Patterson and Katz 1992). In these applications, the amputee must learn to translate and utilize feedback that is not physiologically relevant, for instance, to interpret pressure applied to the upper arm or back as grip force exerted by the hand. Furthermore, different feedback modalities, such as mechanical vibration or electrical input to the skin, are substituted for grip force. The difficulty of integrating this type of feedback is likely a significant factor in the failure of these systems to reach commercial application, even though research on sensory substitution has been ongoing for more than 40 years.

Physiologically relevant sensory feedback is necessary for natural control of prosthetic limbs. The most promising methods for providing this feedback and circumventing the difficulties associated with sensory substitution are neural-machine and brain-machine interfaces. Important attempts have been made in nonhuman-primate models to create sensory-type feedback by stimulation of the sensory cortex (Romo et al. 2000). Recently, a sensory-feedback-mediated increase in limb control performance has been demonstrated in brain-machine interfaces (Suminski et al. 2010; O'Doherty et al. 2011). Here we explore the potential of transfer sensation as a neural-machine interface that provides physiologically relevant sensory feedback from prosthetic limbs.

8.2 Transfer Sensation in Targeted Reinnervation Amputees

Transfer sensation arose unexpectedly in the first targeted reinnervation patient. This was likely due to the denervation of the skin overlying the target muscles through the removal of subcutaneous fat in an effort to improve the quality of electromyographic (EMG) signals (see Chapter 3). It appears that the sensory afferents of transferred nerves grew through the muscle and into the overlying skin, reestablishing functional connections with remaining end organs. The utility of physiologically relevant sensory feedback was quickly recognized, and sensory reinnervation of the target skin was encouraged in some subsequent targeted reinnervation surgeries through *targeted sensory reinnervation*. In this procedure, a cutaneous sensory nerve near the transfer site is cut and the distal end typically coapted end-to-side with the transferred mixed nerve. This allows transferred afferents to reinnervate the now denervated target skin and establish sensation that projects to the missing hand and limb. Transfer sensation is detectable following a period of nerve regeneration (approximately 4–5 months) and is typically very clear and well defined.

Transfer sensation is reported to be qualitatively very different than phantom limb sensation, which is the perception of location, movement, stimulation, and pain in the missing limb and is often experienced by amputees. Phantom sensation is diffuse and difficult to localize; in contrast, transfer sensation is often felt as being precisely located to discrete areas on the missing limb. Transfer sensation arises from the reactivation of amputated and transferred sensory afferents; phantom sensation, however, likely results from changes in functional connectivity and plastic representational reorganization within the central nervous system (Ramachandran et al. 1992; Borsook et al. 1998; Flor et al. 1998; Flor et al. 2000).

8.2.1 Characterization of Transfer Sensation

8.2.1.1 Percept Mapping of Transfer Sensation

After a second established instance of transfer sensation (Kuiken et al. 2007b), it was evident that a foundational understanding of the organization and characteristics of the reinnervated skin was needed. To begin to understand the basic organizational features of transfer sensation, a systematic method was used to map the referred sensations in the reinnervated skin (Kuiken et al. 2007a). A grid of points was placed on the reinnervated skin of each amputee, and the skin at each point was pressed with a set force. After being touched, the amputee reported the location of the resulting projected sensation by referencing a schematic drawing of their missing limb and hand. This created a map of the sensory percepts of the reinnervated target skin (Figure 8.1).

There were many areas on the reinnervated skin where touch only elicited projected sensation to the missing limb. These areas were typically surrounded by an area of skin on which both native chest and projected hand sensation were felt simultaneously (Figure 8.2). The organization of the sensory percepts associated with the reinnervated skin was closely correlated to the regional distribution of the underlying nerve transfers. In many of the targeted sensory reinnervation amputees, the sensory percepts elicited at each of the tested points projected to large areas of the hand. These sensations often encompassed the entire cutaneous distribution of the underlying transferred nerve (see Figure 8.1). Interestingly, aspects of this organizational feature were evident in a later cortical mapping study (see Section 8.2.2).

8.2.1.2 Sensitivity of Reinnervated Skin

Many of the modalities and functions of transfer sensation reflect those of normal skin (Kuiken et al. 2007a). Pressure sensitivity on reinnervated skin approaches that of the hand or chest. Touch threshold sensitivity tests using

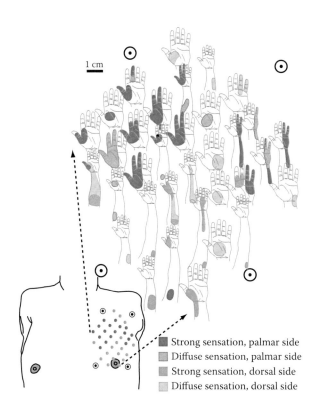

Strong sensation, palmar side
Diffuse sensation, palmar side
Strong sensation, dorsal side
Diffuse sensation, dorsal side

FIGURE 8.1
(See color insert.) The reinnervated chest skin of a shoulder disarticulation TMR patient showing sensations referred to the missing limb elicited by indentation of the reinnervated skin by
a cotton-tipped probe (300 g applied force). (From Kuiken et al., *Proc Natl Acad Sci U S A* 104
(50):20061–20066, 2007. With permission.)

Semmes-Weinstein monofilaments revealed core areas of reinnervated skin
in two shoulder disarticulation amputees that had light touch thresholds of
less than 2 grams of applied force. In certain discrete areas, these thresholds
were between 0.6 and 1.4 grams of applied force, very close to the light touch
thresholds for the palm of the hand (Semmes et al. 1960; Weinstein 1968;
Voerman et al. 1999) (see Figure 8.2). In addition to having near-normal light
touch thresholds, the targeted sensory reinnervation amputees were also
able to discriminate differences in applied force at near-normal levels for the
fingertip (Kotani et al. 2007). These results suggest that the relative pressure
sensitivity of the afferents is maintained in reinnervated skin.

   The reinnervated target skin in these amputees also appears to have thermosensation thresholds reflective of normally innervated hand skin (Kuiken
et al. 2007a). Thresholds were measured by clinical quantitative sensory testing of regions of reinnervated chest skin where sensation was only projected
to the missing hand. Sensation thresholds for cold, warm, cold pain, and heat

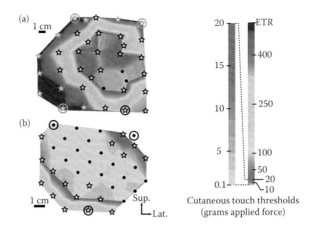

FIGURE 8.2

(See color insert.) Contour plots showing the average amount of force required for two patients (a and b) to feel touch to the reinnervated chest projected to the missing limb. Black dots indicate points where only the missing limb was felt. Stars indicate points where chest was felt at lower force thresholds and the missing limb was felt at higher thresholds. (From Kuiken et al., *Proc Natl Acad Sci U S A* 104 (50):20061–20066, 2007. With permission.)

pain on reinnervated skin fell within normal ranges for the hand (Verdugo and Ochoa 1992; Defrin et al. 2002). In addition, each different thermal sensation was projected to the missing hand.

Electrical stimulation of the reinnervated target skin provided similar results (Kuiken et al. 2007a). As with the thermal and pressure stimuli, electrical stimulation of reinnervated skin generated percepts that were projected to the missing hand (Figure 8.3). Various intensities of electrical stimulation applied to the reinnervated skin demonstrated that a range of A-beta, A-delta, and C-fiber afferents were present. Furthermore, as the intensity of electrical stimulation increased, the surface area of the referred sensations increased—this was likely due to current spread. Pain sensations were felt in the chest and were also projected to the missing limb.

8.2.1.3 Mechanoreceptor Complement in Reinnervated Skin

In an effort to understand the terminal receptor makeup of the reinnervated skin, sensitivity to vibration stimuli was examined on three targeted reinnervation amputees. The relative sensitivity of skin to different frequencies of vibration is indicative of the mechanoreceptors present in the skin. Each mechanoreceptor type exhibits a preference for activation within different vibratory stimulus frequency ranges. For example, in glabrous (hairless) skin, Merkel cells typically respond to 1 to 4 Hz, Meissner's corpuscles respond to 5 to 30 Hz, and Pacinian corpuscles are most responsive to 250 Hz (Bolanowski et al. 1988; Bolanowski et al. 1994). Thus specific frequencies (5, 30, 250, and 400

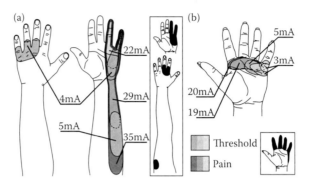

FIGURE 8.3
(See color insert.) Projected sensations elicited by electrical stimulation in two targeted rein-nervation amputees (a and b). Projected fields in blue indicate sensation (threshold) levels, and those in red indicate painful stimulation levels. Insets: Composites of mechanosensory pro-jected fields (300 g applied force) at the same positions/points where the electrical stimulation electrodes were placed. (From Kuiken et al., *Proc Natl Acad Sci U S A* 104 (50):20061–20066, 2007. With permission.)

Hz) were used to determine which receptors might be present. Thresholds were measured on the contralateral chest and arm skin of three shoulder-dis-articulation amputees (one bilateral, two unilateral), as well as on the chest and arm skin of a control population (Schultz et al. 2009). Vibratory stimuli were applied to areas of the reinnervated skin where the amputees could feel only projected hand sensation and no native chest sensation. Vibrotactile frequen-cies were applied at varying amplitudes to determine detection thresholds for each frequency (Kaernbach 1990). The profiles of the response curves for the amputees' reinnervated skin were found to be slightly higher, but similar in shape, to the response profiles for chest skin of non-amputee controls (Schultz et al. 2009) (Figure 8.4). This indicated that a normal complement of recep-tors—Merkel cells, Meissner's corpuscles, and Pacinian corpuscles—had been reinnervated. In addition, the location of perceived limb sensation changed depending on the frequency of vibration, also suggesting that different fre-quencies were detected by different sensory receptor subtypes in the rein-nervated skin (see Figure 8.4). In the two unilateral amputees, the vibration thresholds at multiple locations on the contralateral intact limb were mea-sured. This provided the opportunity to establish an internal control for each amputee and to compare the responses at different locations to those of the non-amputee control population. An important finding was that the vibration detection thresholds for reinnervated skin were more in line with those for hairy (chest) skin than for glabrous (hand) skin. This suggests that the rein-nervated skin environment has a greater degree of influence on the sensitivity of the receptor terminal than the reinnervating afferents.

In addition to the psychophysical characterization of the terminal recep-tors in the human amputees, an electrophysiologic characterization was

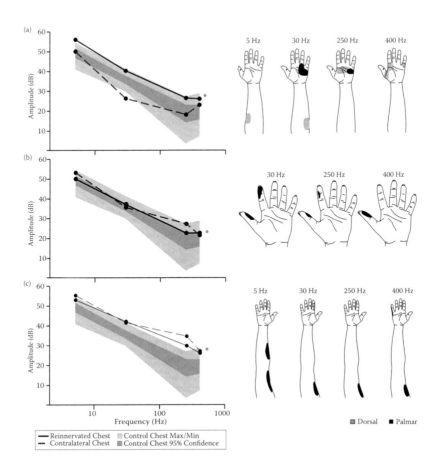

FIGURE 8.4

Left: Vibrotactile detection thresholds on reinnervated and contralateral chests of three targeted reinnervation amputees (a, b, and c). Vibration detection thresholds were measured with a 9-mm probe on both reinnervated and contralateral chest sides. Maximum range (light gray) and 95% confidence intervals (dark gray) of the control population are included for reference. Stars (*) denote potential underestimates where threshold measures are equal to the maximum frequency range of the stimulator. *Right*: Locations of perceived vibrations in the missing limbs of the three amputees following the application of vibration stimuli to the reinnervated chest. (Reprinted and adapted from Schultz et al., Vibrotactile detection thresholds for chest skin of amputees following targeted reinnervation surgery, *Brain Res* 1251:121–129. © [2009]. With permission from Elsevier.)

performed in a targeted reinnervation animal (rat) model to gain a better understanding of terminal receptor functionality. Electrophysiologic recordings were made of activity in transferred sensory afferents. After nerve transfer to denervated muscle and skin, axonal recordings revealed normal response profiles for the reinnervated touch receptors. A mix of slowly adapting and rapidly adapting responses was found, and two specific types of terminal receptors were evident (Figure 8.5). The responses of Pacinian

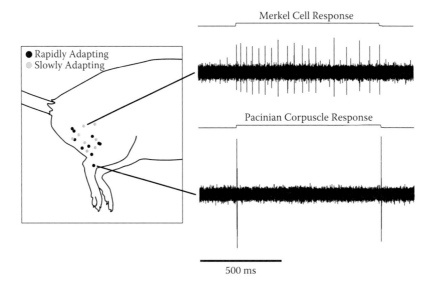

FIGURE 8.5

Left: A composite schematic of receptive field locations from six targeted reinnervation study animals showing an even distribution of rapidly and slowly adapting neural responses. *Right*: Two representative electrophysiological recording traces. Response property profiles were indicative of Merkel cell receptors and Pacinian corpuscle receptors and showed normal dynamic and static response characteristics to input stimuli.

corpuscles and Merkel cells to mechanical inputs on the reinnervated skin appeared to be normal. The presence of normal receptor profiles in reinnervated skin in the animal model matched well with the observations from human psychophysical experiments. The results from this study suggest that regenerating nerves were able to find and reinnervate the appropriate denervated terminal receptor organs.

8.2.1.4 Tactile Acuity of Reinnervated Skin

Psychophysical experiments were undertaken to examine the tactile acuity of the reinnervated skin in humans with targeted sensory reinnervation.

The grating orientation threshold test was specifically designed to study sensory function following nerve injury (Van Boven and Johnson 1994; Bara-Jimenez et al. 2000; Zeuner and Hallett 2003; Walsh et al. 2007). Two particularly important features of the grating orientation threshold test are that (1) it has been validated by electrophysiologic methods and (2) the threshold results correlate better with the sensory deficits reported by subjects than do the results of typical clinical sensory measures (Van Boven and Johnson 1994). The test employs a series of spatial discrimination domes with a series of square wave–shaped grooves cut into the surface. Each dome contains grooves with a different width. These domes are pressed into the skin with

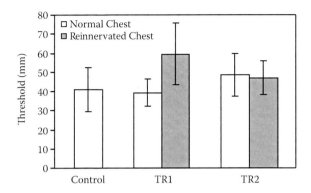

FIGURE 8.6

Grating orientation thresholds for the reinnervated (gray) and normal chest (white) for two targeted reinnervation amputees and the control population ($n = 15$). Error bars indicate 1 SD. This test employed a two-interval forced-choice staircase routine to converge at a threshold value corresponding to 70.7% correct responses. (From Marasco et al., Sensory capacity of reinnervated skin after redirection of amputated upper limb nerves to the chest, *Brain* 132:1441–1448, 2009. With permission from Oxford University Press.)

the grooves oriented either vertically or horizontally, and subjects are asked to tell the experimenter which orientation they perceive. The smallest width of grooves that still allows subjects to reliably detect the orientation is considered the detection threshold width. Importantly, this threshold distance is indicative of the spacing between Merkel cell receptors in the skin being tested (Phillips and Johnson 1981).

Spatial discrimination domes were applied to reinnervated chest skin and contralateral normal chest skin on two targeted reinnervation amputees (Marasco et al. 2009). The threshold values for the reinnervated skin were similar to those for contralateral normal skin (Figure 8.6). This was particularly exciting for two reasons. First, it suggested that even though the reinnervation of the sensory receptors was likely random, their function was normal: the brain appeared to functionally reorganize to compensate for the disorganized somatotopic organization of the skin, building a cohesive sensory picture out of input from randomly innervated skin sensory receptors even in the absence of any type of training. Second, similar threshold values between the reinnervated and normal sides of the chest suggested that the receptor density of the reinnervated skin reflected the normal lower receptor density of the chest and not the higher density of the hand. This observation is important to note in relation to point localization threshold testing (described below).

The point localization test is typically employed to study tactile acuity and hand function after brain injury (Travieso and Lederman 2007; Valentini et al. 2008). For this test, a Y-shaped grid is drawn on the skin and a Semmes-Weinstein monofilament is used to press the skin at the center of the grid and at different distances from the center. The purpose of the test is to determine how far from the center the touch must be for the subject to perceive that the touch occurred in two different places. In these tests, touch stimuli were applied to both the reinnervated and the contralateral chest skin of the amputees (Marasco et al. 2009). Surprisingly, the amputees were able to localize touch stimuli much more effectively on their reinnervated skin than on their contralateral skin (Figure 8.7). The reinnervated chest skin had a similar tactile acuity to the skin of the palm. This result was particularly significant because it suggested that reconnecting afferents from the highly behaviorally relevant hand area of the brain to a new, less behaviorally important skin site conferred an increase in the tactile acuity of the reinnervated skin. This suggests that the processing power of the brain has a considerable impact on tactile acuity above and beyond the receptor density of the skin. Similar observations have been made in the visual system, where the central visual field is more acute than would be predicted by receptor density alone (Azzopardi and Cowey 1993). In contrast, the palm of the hand has a lower tactile acuity than would be predicted from receptor density, probably because of the reduced behavioral importance of this part of the hand in comparison to, for example, the fingertips (Craig and Lyle 2002).

8.2.2 Cortical Mapping of Transfer Sensation

8.2.2.1 Rat Electrophysiologic Studies

Brain mapping studies were conducted in a targeted reinnervation animal model to examine the effects of targeted sensory reinnervation on cortical organization (Marasco and Kuiken 2010). Forelimb amputations were carried out on rats. In the experimental group, the median nerve was transferred to the denervated triceps and denervated forelimb skin. The ulnar and radial nerves were buried in the normally innervated pectoralis muscle to prevent reinnervation. No nerve redirection was carried out in the control group. After at least 13 weeks of recovery, electrophysiologic recordings were made in the sensory cortex of these animals. In the control group, recordings from the area of the sensory cortex devoted to the missing limb (Figure 8.8a) showed no activity: this region was silenced by the amputation (Figure 8.8b). In contrast, this region was active in targeted reinnervation animals (Figure 8.8c). Furthermore, the entire region of the sensory cortex devoted to the limb was reactivated even though only one of the nerves previously serving the limb (the median) was transferred. This suggested that the reactivated median afferents were important to the system and, through changing connections within the sensory cortex (brain plasticity), the processing resources of the previously deactivated cortex were being devoted to the transferred nerve. In addition to the implications for sensory processing, this result suggests that the large sensory percepts reported by human targeted reinnervation amputees may be perceptual manifestations of a similar widespread cortical reactivation.

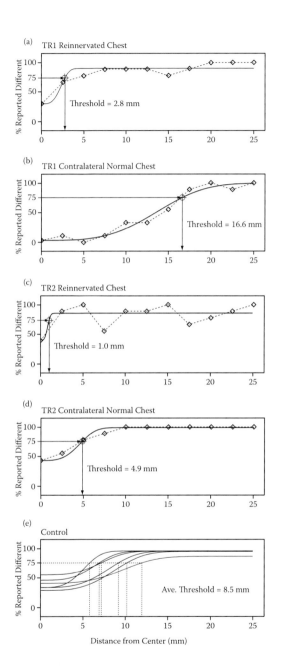

FIGURE 8.7
Point localization thresholds for the reinnervated (a, c) and normal contralateral (b, d) chest for two targeted reinnervation amputees (TR1 and TR2) and six non-amputee controls (e). Touch stimuli were applied with a 10-g Semmes-Weinstein monofilament. A psychometric curve was fit to the data with the 75% difference point taken as the threshold. (From Marasco et al., Sensory capacity of reinnervated skin after redirection of amputated upper limb nerves to the chest, *Brain* 132:1441–1448, 2009. With permission from Oxford University Press.)

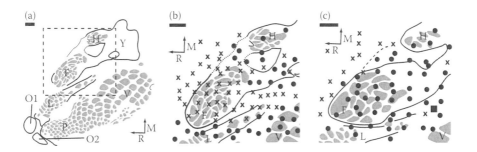

FIGURE 8.8

(a) Complete map showing organization of the primary somatosensory cortex (S1) in the rat. F, Forepaw barrel subfield; H, hindpaw barrel subfield; Y, body; V, vibrissae; C, chin/lower lip; O1, oral module 1; O2, oral module 2; L, lower lip; P, buccal pad; R, rostral; M, medial. Gray spots delineate cytochrome oxidase–stained barrel patterns within S1, and black lines define the borders of the S1 representation. Dashed square shows the area of the forelimb barrel subfield that is the focus of center and right panels. Barrel patterns within the forelimb barrel subfield correspond to digits of the forepaw. (b) Representative result of microelectrode recordings in S1 of the left hemisphere of a control rat after forelimb amputation. The forelimb barrel subfield was found to be largely silent in these animals. (c) Representative result of microelectrode recordings in S1 of the left hemisphere of a rat after forelimb amputation and TMR. The forelimb barrel subfield was found to be largely active in these animals, although only the median nerve was transferred to the target skin and muscle. Scale bars, 500 μm. ●, Responsive site; X, unresponsive site; ■, dual receptive field site. (Modified from Marasco and Kuiken, *J Neurosci*, 30 (47):16008–16014, 2010. With permission.)

The widespread reactivation of the sensory cortex observed in the cortical mapping experiments in rats, along with the reduced density of reinnervated receptors indicated by human psychophysical studies, suggested that changes in brain connectivity occur in response to targeted reinnervation surgery. An additional experiment with this targeted reinnervation animal model was used to provide a glimpse into the possible changes in cortical circuitry. This experiment involved measuring somatosensory evoked potentials during cortical recording. Evoked potentials are electrical signals recorded from the nervous system in response to physical stimuli. In this experiment, evoked potentials were recorded from the sensory cortex in response to activation of cutaneous touch receptors by applying a small electrical shock at the skin surface. The elapsed time between the application of the shock and the observation of evoked potentials (i.e., the time taken for the response to travel to the brain) is the evoked potential response latency. Evoked potential latencies of around 7 ms indicate a direct neural connection to the cortex. In contrast, evoked potentials of 10 ms or greater indicate an indirect connection via multistep lateral pathways involving a number of synaptic connections between brain neurons. This experiment demonstrated that some transferred axons had a direct, short-latency connection from the skin to the sensory cortex area representing the missing limb, whereas other transferred axons were connected via indirect, lateral pathways (Figure 8.9). From a wider perspective, this

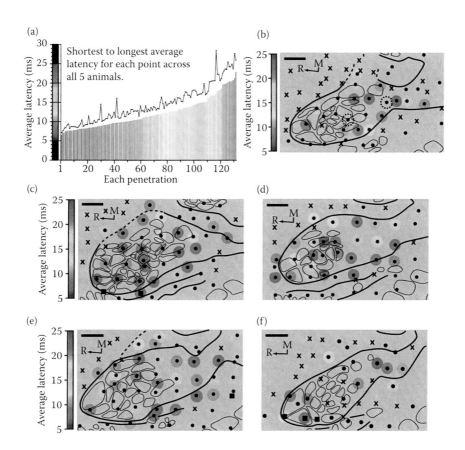

FIGURE 8.9

(See color insert.) Latency measurements for electrically evoked somatosensory potentials from stimulus artifact to cortical response from the reinnervated target skin to the electrode penetration points within the forelimb barrel subfield of five targeted reinnervation rats. (a) Average latency for each penetration (*n* = 132) across all recording cases, arranged from shortest to longest. Black line indicates SD for each latency. (b–f) Latency measurements for each electrode penetration in the forelimb barrel subfield overlaid with cytochrome oxidase-delineated digit barrel structures. Colored circles indicate latency in milliseconds; (●) indicates a responsive site, (X) an unresponsive site, (■) a dual receptive field site; R, rostral; M, medial. Dashed circles indicate inability to measure average latency. For clarity, the diameter of the colored latency points is set to reflect 1.5 times the 100-μm hypothetical recording sphere at the electrode tip (Robinson 1968). Average latencies for each penetration range from 6.6 ms to 23.0 ms. Often, shorter latencies lie directly adjacent to longer latencies. In (f), no latencies were recorded at four points within the forelimb barrel subfield. Scale bars represent 500 μm. (From Marasco and Kuiken, *J Neurosci,* 30 (47):16008–16014, 2010. With permission.)

suggested that there were a limited number of transferred afferents connected directly to the forelimb representation and that the reactivation of the entire region of the sensory cortex corresponding to the amputated limb likely reflected an increase in processing power for these afferents mediated by brain reorganization.

8.2.2.2 Human High-Density Electroencephalography Studies

A brain mapping study using high-density electroencephalography (EEG) was undertaken to examine sensory cortical remapping in an amputee with targeted sensory reinnervation (Yao et al. 2011). This individual's median, ulnar, and radial nerves were transferred to his medial biceps, brachialis, and lateral triceps, respectively. Somatosensory evoked potentials were recorded during electrical stimulation of an area of the reinnervated skin that produced the perception of sensation on the tip of his missing middle finger. EEG experiments were conducted immediately prior to targeted reinnervation surgery and then at 6, 12, and 24 months after the procedure. Prior to targeted reinnervation, EEG revealed a diffuse activity pattern in the sensory cortex that was distributed over both sides of the brain. This bilateral activation has also been demonstrated in lower limb amputees and is likely a result of central changes brought on by reductions in local inhibition and structural changes within connected pathways (Simoes et al. 2012). In the targeted reinnervation patient, this bilateral activation was also evident 6 months and 12 months after targeted reinnervation. However, at 24 months, the cortical activity in the primary somatosensory cortex was concentrated in the contralateral hemisphere. This more normal spatial distribution of activity likely reflected a return toward the preinjury representational mapping as a result of the reinnervation by the amputated nerves.

8.3 Application of Transfer Sensation for Sensory Feedback

8.3.1 Current Uses: Haptic Tactors

After nerve transfer, the reinnervated mechanoreceptors in reinnervated skin appear to function normally; thus mechanical input can be applied to the skin as a means of providing feedback. A potentially useful approach for providing pressure and vibration input is through miniature haptic displays or *tactors*—robotic devices that can provide mechanical stimulation of the reinnervated skin in proportion to touch input measured by a sensor on the prosthetic hand (Kim et al. 2007a; Kim et al. 2007b; Kim et al. 2010). With this approach, it is possible to stimulate areas of the reinnervated skin that are

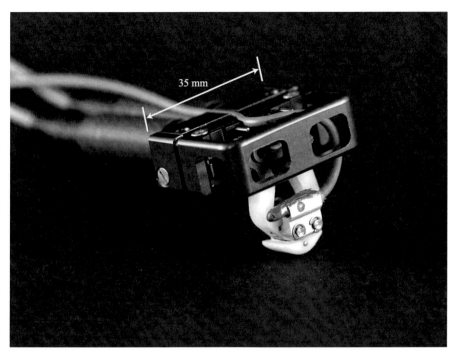

FIGURE 8.10
Six-bar mechanism with gear constraint and 10-mm brushed motor (G10). Multifunction tactors such as the G10 can provide the sensations of contact, pressure, vibration, shear force, and temperature. (Photograph provided courtesy of Kinea Design, LLC.)

perceptually related to the location of the sensor on the prosthetic hand; in effect, creating a physiologically relevant artificial sense of touch. A recent lightweight, wearable tactor uses a six-bar skewed parallelogram approach that is able to provide up to 12 mm of excursion and apply up to 9 N of force (Kim et al. 2010) (Figure 8.10). This device can also provide a tapping sensation (for goal confirmation) in addition to static and dynamic pressure, vibration, and shear forces. In addition to providing mechanosensory input, a subset of tactor devices can provide thermal input to the skin (Kim et al. 2007a). Haptic tactors used in conjunction with TMR-controlled prosthetic limbs hold the promise of providing a truly bidirectional user interface for artificial arms. However, progress in developing clinically robust systems has been slow for two reasons. First, the reinnervated skin is usually over the reinnervated muscle. Thus tactors must share the skin surface with EMG electrodes. This space is limited and the tactors can introduce significant noise into the EMG signal. Second, tactors are usually mounted in prosthetic sockets, and it is challenging both to accommodate the bulk of the tactor and to have the tactor track skin movement and apply a repeatable stimulus as the socket moves during use.

8.3.1.1 Force and Vibratory Feedback

As described earlier, it is possible to provide amputees who have undergone targeted sensory reinnervation with vibration and force feedback. This offers the possibility that pressure and vibratory feedback could also be applied through a tactor system to provide functional pressure and texture discrimination. One of the most useful cutaneous touch feedback modes that could be restored is the sense of goal confirmation. This would allow the amputee to feel when contact with an object is made and not have to rely exclusively on visual monitoring. A sense of grip force is also highly important. The sensory feedback system could be utilized to give the amputee an intuitive sense of how hard they were squeezing an object without having to rely on the sound of the prosthesis motor or visual cues. Vibrotactile input from a prosthesis may also provide texture discrimination (Hollins et al. 2002).

8.3.1.2 Embodiment of the Prosthetic Limb

In addition to providing direct modes of haptic feedback from a prosthesis, transfer sensation provides a physiologically relevant avenue for tapping into cognitive mechanisms of limb embodiment. Cortical mapping experiments in the TMR animal model suggested that reinnervated skin was connected directly to the region of the brain dedicated to the missing limb. In addition, the widespread reactivation of the processing area in the sensory cortex associated with the missing limb suggested that inputs from transferred sensory nerves are given a certain amount of privilege in the brain. These results are well aligned with the evidence of increased tactile processing observed in the human psychophysical grating orientation and point localization experiments. In these amputees, it appeared that if the hand sensation was strongly physiologically relevant, the resulting perceptual effect might be leveraged to make amputees feel as if their prosthesis was an integral part of their body.

The sense of body ownership is intrinsically linked to vision and the sense of touch. In a perceptual illusion called the "rubber hand illusion," mechanisms of visual-tactile integration are used to fool subjects into thinking that a rubber hand, positioned in place of their normal hand, is a part of their body (Figure 8.11). This is accomplished by applying stimuli to the rubber hand, which the subject can see, while simultaneously applying the same stimuli at the same location on the normal arm, which the subject cannot see. The direct correlation of stimulus locations on the rubber hand and normal arm is analogous to providing physiologically relevant feedback from a prosthesis (e.g., touching the prosthetic fingers and applying a stimulus to a region in which amputees perceive touch to their missing fingers). A number of lines of evidence suggested that transfer sensation is highly physiologically relevant: (1) the multiple classes of the reinnervated cutaneous sensory receptors were functional (Kuiken et al. 2007a); (2) the sensation from the

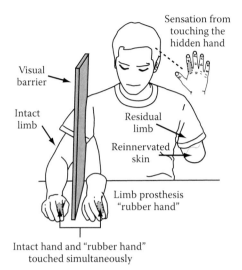

Sensation from
touching the
hidden hand

Visual
barrier

Intact
limb

Residual
limb

Reinnervated
skin

Limb prosthesis
"rubber hand"

Intact hand and "rubber hand"
touched simultaneously

FIGURE 8.11
Schematic diagram of the experimental setup for the rubber hand illusion using the contralateral intact hand of the amputee. To generate the illusion, the amputee watched as the
investigator repeatedly touched the amputee's hidden intact hand at different locations while
simultaneously touching the corresponding locations on the visible rubber hand. (Adapted
from Marasco et al., Robotic touch shifts perception of embodiment to a prosthesis in targeted
reinnervation amputees, *Brain* 134 (3):747–758, 2011. With permission from Oxford University
Press.)

transferred nerves travels through the sensory channels of the missing limb
(Kuiken et al. 2007a; Marasco and Kuiken 2010); (3) the sensory input from
the target skin appeared to be processed in appropriate parts of the brain
(Marasco and Kuiken 2010); (4) the nerves serving the reinnervated target
skin appeared to be important to the system and were given an expanded
representation in the sensory cortex (Marasco and Kuiken 2010); and (5) tactile acuity of the reinnervated target skin appeared to be higher than that of
contralateral control skin. Given these interesting outcomes, it seemed likely
that a physiologically relevant sense of touch in a prosthetic limb could be
linked with visual cues to modulate the sense of prosthetic limb ownership
in a targeted reinnervation amputee.

A variation of the rubber hand illusion was used in a series of experiments
to determine if a physiologically relevant touch interface could create a perceptual sense of ownership of a prosthetic limb (Marasco et al. 2011). A tactor
was placed on the reinnervated skin of two targeted reinnervation amputees and a load cell was placed on the prosthesis (Figure 8.12). Touch was
applied to the load cell in combination with different visual cues to develop
an understanding of the amputees' perception of different visual/tactile
conditions. Touch input from the robotic tactor was always felt in the fingertips and hand. In some conditions, the amputees watched the prosthesis

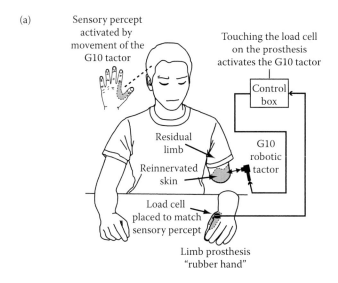

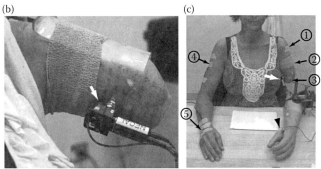

FIGURE 8.12

(a) Schematic diagram of the experimental setup for the rubber hand illusion; a G10 tactor was used to provide physiologically relevant touch feedback. The illusion was generated on the amputated side of subjects by having them watch the investigator touch the prosthetic hand and load cell while the G10 tactor pressed into the reinnervated target skin of the residual limb. (b) The placement of the G10 tactor on the reinnervated skin of a TMR amputee. The plunger (white arrow) pushes into a region of skin where the subject feels sensation projected to the dorsal skin between digits 1 and 2 of her missing hand. (c) The placement of the prosthetic limb on the table in front of the subject. The subject positioned the arm to where it looked most natural. The G10 tactor can be seen on the inner aspect of her residual limb (white arrow), and the load cell that provides touch input to the G10 tactor can be seen placed in the center of the projected field of sensation (black arrowhead). Numbers mark the location of each thermistor: (1) proximal residual limb, (2) mid residual limb, (3) distal residual limb, (4) proximal intact limb, and (5) intact hand. (Adapted from Marasco et al., Robotic touch shifts perception of embodiment to a prosthesis in targeted reinnervation amputees, *Brain* 134 (3):747–758, 2011. With permission from Oxford University Press.)

being touched on the hand in the place they felt the touch occurring on their missing hand (i.e., visual and tactile input was matched). In other conditions, the amputees watched the prosthesis being touched in a place that was different from where they felt touch on their missing hand (i.e., visual and tactile input was mismatched). Their perceptual experience was measured through three independent experimental approaches: questionnaires, psychophysical temporal order judgments, and objective measurements of limb temperature.

In all three experimental approaches, the amputees showed signs of perceptual integration of the prosthetic limb into their self-image when visual input aligned with touch input. In other words, when the amputees observed a match between what they were seeing and what they were feeling, they involuntarily took perceptual ownership of the prosthesis. In questionnaires given after experiments it was evident that when tactile and visual inputs were matched, the amputees more strongly agreed with terms suggesting that the prosthesis was part of their body (Figure 8.13). Moreover, when visual input matched touch input, psychophysical temporal order judgments revealed a side-specific change in how the amputees cognitively processed vibratory input to each limb (Figure 8.14). Furthermore, residual limb temperature increased when visual and tactile inputs were in alignment. The distal residual limbs of the amputees in this study were initially typically cooler than their intact limbs. The observed increase in residual limb temperature may reflect a renewed metabolic commitment to the missing limb and an incorporation of the prosthesis into the self-image of the amputee (Armel and Ramachandran 2003) (Figure 8.15).

The results suggest that this strong perceptual effect can be driven entirely by a relatively simple robotic touch interface that could operate as an independent on-board feature of a prosthetic limb. Furthermore, the long-term use of a physiologically relevant touch interface may help to bolster mechanisms of functional motor control and may help amputees to feel that their prosthetic limb is an integral part of their body instead of simply a tool that they wear.

8.3.2 Potential Benefits and Future Use of Targeted Sensory Reinnervation

The studies described previously have demonstrated that the use of a single tactor effectively provides the amputee with graded pressure sensation, vibration detection (including contact information), and a sense of cognitive investment in the limb. The successful integration of these approaches into prosthetic limbs could greatly benefit amputees, particularly those with median nerve transfer sensation providing thumb and index finger feedback. This would allow the amputee to experience useful touch input, such as when they touch an object (goal confirmation) or how hard they are squeezing an object (graded pressure). In addition, this

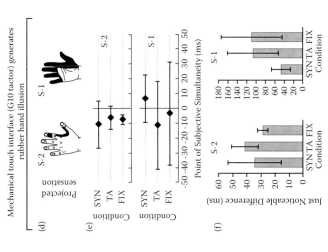

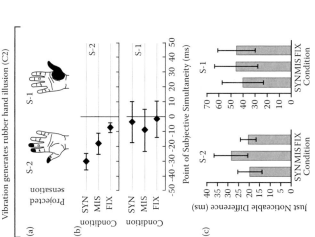

FIGURE 8.13

Results of temporal order judgment tasks for the two different experimental configurations each with three different stimulus conditions (error bars = 95% confidence interval). (a–c) Use of the vibratory units to generate the rubber hand illusion; (d–f) use of the touch interface to generate the rubber hand illusion while vibratory input for the temporal order judgment task was applied to the shoulders. (a) Diagrams of projected sensation elicited by the vibratory input applied to the reinnervated skin. (b) The point of subjective simultaneity. (c) The just noticeable difference. (d) Diagrams of projected sensation elicited by the G10 tactor pushing into the reinnervated skin. (e) The point of subjective simultaneity. (f) The just noticeable difference. FIX = fixation; MIS = spatial mismatch; SYN = synchronous; TA = temporal asynchrony. (Adapted from Marasco et al., Robotic touch shifts perception of embodiment to a prosthesis in targeted reinnervation amputees, *Brain* 134 (3):747–758, 2011. With permission from Oxford University Press.)

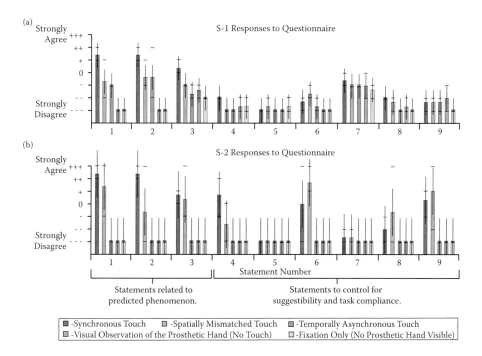

FIGURE 8.14

(See color insert.) Questionnaire results for TMR subject S-1 (a) and TMR subject S-2 (b) for five different stimulus conditions. Vertical error bars indicate 95% confidence intervals (S-1 = ± 0.975, S-2 = ± 1.865) from a multiple comparisons procedure; horizontal lines indicate range, n = 3. Significance was judged by non-overlap of confidence intervals. Statements 1 through 3 were predicted phenomena: question #1, "I felt the touch of the investigator on the prosthetic hand"; 2, "It seemed as if the investigator caused the touch sensations that I was experiencing"; 3, "It felt as if the prosthetic hand was my hand." Statements 4 through 9 were controls: 4, "It felt as if my residual limb was moving towards the prosthetic hand"; 5, "It felt as if I had three arms"; 6, "I could sense the touch of the investigator somewhere between my residual limb and the prosthetic hand"; 7, "My residual limb began to feel rubbery"; 8, "It was almost as if I could see the prosthesis moving towards my residual limb"; 9, "The prosthesis started to change shape, color and appearance so that it started to visually resemble the residual limb." For S-2, confidence intervals for scores related to ownership of the limb did not overlap with control scores. There was some overlap between the ownership and control scores for S-1. (Adapted from Marasco et al., Robotic touch shifts perception of embodiment to a prosthesis in targeted reinnervation amputees, *Brain* 134 (3):747–758, 2011. With permission from Oxford University Press.)

type of interface could provide information on surface texture (vibration detection) and temperature (thermosensation). As increasingly sophisticated prosthetic limb components become available, future approaches to sensory feedback could include sensors that detect touch input from multiple digits and the palm of the hand. We anticipate that future devices will utilize the rich array of touch feedback made available with targeted sensory reinnervation.

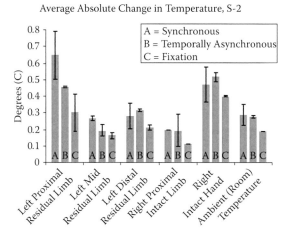

Average Absolute Change in Temperature, S-2

FIGURE 8.15

Average absolute changes in skin temperature calculated from the onset of the stimulus condition (time = 1 min) to the termination of the experiment (time = 17 min) for subject S-2 at three points on her residual limb, two points on her intact limb, and for the ambient room temperature measured during the three different stimulus conditions (error bars = ±1SD, $n = 2$). (Adapted from Marasco et al., Robotic touch shifts perception of embodiment to a prosthesis in targeted reinnervation amputees, *Brain* 134 (3):747–758, 2011. With permission from Oxford University Press.)

8.3.2.1 Leverage of Plasticity Effects

The studies described earlier also demonstrate that the brain has a high degree of intrinsic flexibility. Furthermore, the system appears to be primed to make the most of physiologically relevant sensory feedback. As understanding of the neuromechanistic aspects of changes that occur following nerve redirection increases, there appears to be great promise in the ability to maximize the effects of brain plasticity and nerve regeneration. Long-term use of sensory feedback would certainly continue to refine the mechanisms of plasticity to make the most effective use of the restored sensory input.

8.3.2.2 Potentiation of Motor Control

Motor control is most effective in combination with input from the sensory system. Recent work has demonstrated that the addition of sensory feedback to a brain-machine interface potentiates motor control (Suminski et al. 2010). We anticipate that as the implementation of bidirectional interfaces progresses, we will realize many benefits associated with the restoration of natural closed-loop control, producing further improvement in function of prosthetic devices for people with limb loss.

References

Armel, K. C., and V. S. Ramachandran. 2003. Projecting sensations to external objects: evidence from skin conductance response. *Proc Biol Sci* 270 (1523):1499–1506.

Atkins, D., D. Heard, and W. Donovan. 1996. Epidemiologic overview of individuals with upper-limb loss and their reported research priorities. *J Prosthet Orthot* 8:2–11.

Azzopardi, P., and A. Cowey. 1993. Preferential representation of the fovea in the primary visual cortex. *Nature* 361 (6414):719–721.

Bara-Jimenez, W., P. Shelton, and M. Hallett. 2000. Spatial discrimination is abnormal in focal hand dystonia. *Neurology* 55 (12):1869–1873.

Beeker, T. W., J. During, and A. Den Hertog. 1967. Artificial touch in a hand-prosthesis. *Med Biol Eng* 5 (1):47–49.

Biddiss, E., D. Beaton, and T. Chau. 2007. Consumer design priorities for upper limb prosthetics. *Disabil Rehabil Assist Technol* 2 (6):346–357.

Bolanowski, S. J., G. A. Gescheider, and R. T. Verrillo. 1994. Hairy skin: psychophysical channels and their physiological substrates. *Somatosens Mot Res* 11 (3):279–290.

Bolanowski, S. J., Jr., G. A. Gescheider, R. T. Verrillo, and C. M. Checkosky. 1988. Four channels mediate the mechanical aspects of touch. *J Acoust Soc Am* 84 (5):1680–1694.

Borsook, D., L. Becerra, S. Fishman, et al. 1998. Acute plasticity in the human somatosensory cortex following amputation. *Neuroreport* 9 (6):1013–1017.

Craig, J. C., and K. B. Lyle. 2002. A correction and a comment on Craig and Lyle (2001). *Percept Psychophys* 64 (3):504–506.

Defrin, R., A. Ohry, N. Blumen, and G. Urca. 2002. Sensory determinants of thermal pain. *Brain* 125 (Pt 3):501–510.

Ehrsson, H. H., C. Spence, and R. E. Passingham. 2004. That's my hand! Activity in premotor cortex reflects feeling of ownership of a limb. *Science* 305 (5685):875–877.

Ernst, M. O., and M. S. Banks. 2002. Humans integrate visual and haptic information in a statistically optimal fashion. *Nature* 415 (6870):429–433.

Ernst, M. O., M. S. Banks, and H. H. Bulthoff. 2000. Touch can change visual slant perception. *Nat Neurosci* 3 (1):69–73.

Flor, H., T. Elbert, W. Mühlnickel, C. Pantev, C. Wienbruch, and E. Taub. 1998. Cortical reorganization and phantom phenomena in congenital and traumatic upper-extremity amputees. *Exp Brain Res* 119 (2):205–212.

Flor, H., W. Mühlnickel, A. Karl, et al. 2000. A neural substrate for nonpainful phantom limb phenomena. *Neuroreport* 11 (7):1407–1411.

Fourneret, P., J. Paillard, Y. Lamarre, J. Cole, and M. Jeannerod. 2002. Lack of conscious recognition of one's own actions in a haptically deafferented patient. *Neuroreport* 13 (4):541–547.

Gallagher, S., and J. Cole. 1995. Body schema and body image in a deafferented subject. *J Mind Behav* 16:369–390.

Gescheider, G. A., J. H. Wright, and R. T. Verrillo. 2009. *Information-Processing Channels in the Tactile Sensory System: A Psychophysical and Physiological Analysis,* edited by S. W. Link and J. T. Townsend. New York: Psychology Press, Taylor & Francis Group.

Hollins, M., S. J. Bensmaia, and E. A. Roy. 2002. Vibrotaction and texture perception. *Behav Brain Res* 135 (1–2):51–56.

Huang, F. C., R. B. Gillespie, and A. D. Kuo. 2007. Visual and haptic feedback contribute to tuning and online control during object manipulation. *J Mot Behav* 39 (3):179–193.

Jenmalm, P., and R. S. Johansson. 1997. Visual and somatosensory information about object shape control manipulative fingertip forces. *J Neurosci* 17 (11):4486–4499.

Kaernbach, C. 1990. A single-interval adjustment-matrix (SIAM) procedure for unbiased adaptive testing. *J Acoust Soc Am* 88 (6):2645–2655.

Kim, K., J. E. Colgate, and M. A. Peshkin. 2007a. *A pilot study of a thermal display using a miniature tactor for upper extremity prosthesis.* Paper read at Frontiers in the Convergence of Bioscience and Information Technologies, 2007. FBIT

Kim, K., J. E. Colgate, M. A. Peshkin, J. J. Santos-Munne, and A. Makhlin. 2007b. *A miniature tactor design for upper extremity prosthesis.* Paper read at Frontiers in the Convergence of Bioscience and Information Technologies, 2007. FBIT 2007, 11–13 Oct. 2007.

Kim, K., J. E. Colgate, J. J. Santos-Munne, A. Makhlin, and M. A. Peshkin. 2010. On the design of miniature haptic devices for upper extremity prosthetics. *IEEE/ASME Trans Mechatronics* 15 (1):27–39.

Kotani, K., S. Ito, T. Miura, and K. Horii. 2007. Evaluating tactile sensitivity adaptation by measuring the differential threshold of archers. *J Physiol Anthropol* 26 (2):143–148.

Kuiken, T. A., G. A. Dumanian, R. D. Lipschutz, L. A. Miller, and K. A. Stubblefield. 2004. The use of targeted muscle reinnervation for improved myoelectric prosthesis control in a bilateral shoulder disarticulation amputee. *Prosthet Orthot Int* 28 (3):245–253.

Kuiken, T. A., P. D. Marasco, B. A. Lock, R. N. Harden, and J. P. A. Dewald. 2007a. Redirection of cutaneous sensation from the hand to the chest skin of human amputees with targeted reinnervation. *Proc Natl Acad Sci USA* 104 (50):20061–20066.

Kuiken, T. A., L. A. Miller, R. D. Lipschutz, et al. 2007b. Targeted reinnervation for enhanced prosthetic arm function in a woman with a proximal amputation: a case study. *Lancet* 369 (9559):371–380.

Marasco, P. D., K. Kim, J. E. Colgate, M. A. Peshkin, and T. A. Kuiken. 2011. Robotic touch shifts perception of embodiment to a prosthesis in targeted reinnervation amputees. *Brain* 134 (3):747–758.

Marasco, P. D., and T. A. Kuiken. 2010. Amputation with median nerve redirection (targeted reinnervation) reactivates forepaw barrel subfield in rats. *J Neurosci* 30 (47):16008–16014.

Marasco, P. D., A. E. Schultz, and T. A. Kuiken. 2009. Sensory capacity of reinnervated skin after redirection of amputated upper limb nerves to the chest. *Brain* 132:1441–1448.

Mountcastle, V. B. 2005. *The Sensory Hand: Neural Mechanisms of Somatic Sensation.* Cambridge, MA: Harvard University Press.

O'Doherty, J. E., M. A. Lebedev, P. J. Ifft, et al. 2011. Active tactile exploration using a brain-machine-brain interface. *Nature* 479 (7372):228–231.

Patterson, P. E., and J. A. Katz. 1992. Design and evaluation of a sensory feedback system that provides grasping pressure in a myoelectric hand. *J Rehabil Res Dev* 29 (1):1–8.

Phillips, J. R., and K. O. Johnson. 1981. Tactile spatial resolution. II. Neural representation of bars, edges, and gratings in monkey primary afferents. *J Neurophysiol* 46 (6):1192–1203.

Prior, R. E., and J. Lyman. 1975. Electrocutaneous feedback for artificial limbs. Summary progress report. February 1, 1974, through July 31, 1975. *Bull Prosthet Res* (10-24):3–37.

Ramachandran, V. S., D. Rogers-Ramachandran, and M. Stewart. 1992. Perceptual correlates of massive cortical reorganization. *Science* 258 (5085):1159–1160.

Robinson, D. A. 1968. The electrical properties of metal microelectrodes. *Proc IEEE* 56 (6):1065–1071.

Romo, R., A. Hernandez, A. Zainos, C. D. Brody, and L. Lemus. 2000. Sensing without touching: psychophysical performance based on cortical microstimulation. *Neuron* 26 (1):273–278.

Sainburg, R. L., M. F. Ghilardi, H. Poizner, and C. Ghez. 1995. Control of limb dynamics in normal subjects and patients without proprioception. *J Neurophysiol* 73 (2):820–835.

Schultz, A. E., P. D. Marasco, and T. A. Kuiken. 2009. Vibrotactile detection thresholds for chest skin of amputees following targeted reinnervation surgery. *Brain Res* 1251:121–129.

Scott, R. N., R. H. Brittain, R. R. Caldwell, A. B. Cameron, and V. A. Dunfield. 1980. Sensory-feedback system compatible with myoelectric control. *Med Biol Eng Comput* 18 (1):65–69.

Semmes, J., S. Weinstein, L. Ghent, and H. Teuber. 1960. *Somatosensory Changes after Penetrating Brain Wounds in Man*. Cambridge, MA: Harvard University Press.

Simoes, E. L., I. Bramati, E. Rodrigues, et al. 2012. Functional expansion of sensorimotor representation and structural reorganization of callosal connections in lower limb amputees. *J Neurosci* 32 (9):3211–3220.

Sober, S. J., and P. N. Sabes. 2005. Flexible strategies for sensory integration during motor planning. *Nat Neurosci* 8 (4):490–497.

Suminski, A. J., D. C. Tkach, A. H. Fagg, and N. G. Hatsopoulos. 2010. Incorporating feedback from multiple sensory modalities enhances brain-machine interface control. *J Neurosci* 30 (50):16777–16787.

Travieso, D., and S. J. Lederman. 2007. Assessing subclinical tactual deficits in the hand function of diabetic blind persons at risk for peripheral neuropathy. *Arch Phys Med Rehabil* 88 (12):1662–1672.

Valentini, M., U. Kischka, and P. W. Halligan. 2008. Residual haptic sensation following stroke using ipsilateral stimulation. *J Neurol Neurosurg Psychiatry* 79 (3):266–270.

Van Boven, R. W., and K. O. Johnson. 1994. A psychophysical study of the mechanisms of sensory recovery following nerve injury in humans. *Brain* 117 (1):149–167.

Verdugo, R., and J. L. Ochoa. 1992. Quantitative somatosensory thermotest. A key method for functional evaluation of small calibre afferent channels. *Brain* 115 (Pt 3):893–913.

Voerman, V. F., J. van Egmond, and B. J. P. Crul. 1999. Normal values for sensory thresholds in the cervical dermatomes: a critical note on the use of Semmes-Weinstein monofilaments. *Am J Phy Med Rehabil* 78 (1):24–29.

Walsh, R., J. P. O'Dwyer, I. H. Sheikh, S. O'Riordan, T. Lynch, and M. Hutchinson. 2007. Sporadic adult onset dystonia: sensory abnormalities as an endophenotype in unaffected relatives. *J Neurol Neurosurg Psychiatry* 78 (9):980–983.

Weinstein, S. 1968. Intensive and extensive aspects of tactile sensitivity as a function of body part, sex and laterality. In *The Skin Senses*, edited by D. R. Kenshalo. Springfield, IL: Charles C Thomas.

Yao, J., C. Carmona, A. Chen, T. Kuiken, and J. Dewald. 2011. Sensory cortical re-mapping following upper-limb amputation and subsequent targeted reinnervation: a case report. *Conf Proc IEEE Eng Med Biol Soc* 2011:1065–1068.

Zeuner, K. E., and M. Hallett. 2003. Sensory training as treatment for focal hand dystonia: a 1-year follow-up. *Mov Disord* 18 (9):1044–1047.

Zhang, W., J. A. Johnston, M. A. Ross, et al. 2011. Effects of carpal tunnel syndrome on adaptation of multi-digit forces to object weight for whole-hand manipulation. *PLoS One* 6 (11):e27715.

9

Surgical and Functional Outcomes of Targeted Muscle Reinnervation

Laura A. Miller, Kathy A. Stubblefield,
Robert D. Lipschutz, Blair A. Lock, Jason M. Souza,
Gregory A. Dumanian, and Todd A. Kuiken

CONTENTS

9.0 Introduction

Targeted muscle reinnervation (TMR) was initially performed as an experimental procedure designed to improve control of myoelectric prostheses for individuals with amputations at the shoulder disarticulation level and transhumeral level. Beginning in 2002, the first TMR procedures were performed at Northwestern Memorial Hospital (NMH) in Chicago, with prosthetic fitting, training, and extensive testing in a research setting at the Rehabilitation Institute of Chicago (RIC). The use of TMR as an experimental technique was subsequently expanded to include military amputees, through collaboration with San Antonio Military Medical Center (SAMMC) and Walter Reed National Military Medical Center (WRNMMC). TMR is now an established clinical procedure and, to date, has been performed in more than a dozen institutions with more than 60 patients worldwide. Here we review surgical outcomes for initial TMR procedures, performed through February 2012 at NMH and SAMMC, and functional outcomes for the first six patients treated at RIC. All outcomes data were obtained under approved protocols from the Institutional Review Boards of the appropriate institutions.

9.1 Surgical Goals

As outlined in the surgical technique chapter (Chapter 3), the key elements of the TMR procedure involve identification of the residual brachial nerves, mobilization of these donor nerves to attain sufficient length to reach the intended target muscles, division of the recipient muscles into independent functional units based on neurovascular anatomy, and coaptation of the donor nerves to discrete motor end points on the target muscle segments. The overall objective is to create strong, spatially distinct electromyographic (EMG) signals that correspond to the motor intent of the transferred nerves. Intuitive control of a myoelectric prosthesis can then be achieved by pairing EMG signals from reinnervated muscle segments with their correlate prosthetic function. From a surgical standpoint, the TMR procedure was considered successful if EMG signals generated from the reinnervated muscle sites—in response to attempted activation of the transferred nerve—were of usable magnitude for myoelectric prosthesis control.

In the transhumeral amputee, residual limb anatomy is relatively predictable and surgical strategies follow a fairly uniform course. Two myoelectric control sites are typically available (over the biceps and triceps muscles) prior to surgery. The goal is to create two additional EMG control sites by transferring the median nerve and the distal radial nerve to appropriate target muscles—routinely the short head of the biceps and the lateral head of the triceps, respectively. The musculocutaneous innervation of the long head of the biceps and the proximal radial innervation of the long head of the triceps are preserved. In individuals with long residual limbs, it may be possible to create a fifth control site through transfer of the ulnar nerve to the remnant brachialis muscle.

The goal for the shoulder disarticulation amputee is to transfer all four of the main brachial nerves to create four control sites. However, because these proximal amputations are frequently caused by high-energy trauma, the TMR dissection can be complicated by significant anatomic distortion and damage to the donor nerves or potential target muscles. Surgical plans must therefore be individualized according to the remaining anatomy, available target muscles, and condition of the residual nerves.

9.1.1 Patient Population

Between February 2002 and February 2012, TMR procedures were performed on 27 patients at NMH and SAMMC: 10 individuals with shoulder disarticulations and 17 with transhumeral amputations. The age of patients at the time of TMR surgery ranged from 18 to 55 years; age did not have any noticeable effect on surgical outcome. Three of the individuals had bilateral amputations; in all of these individuals, TMR was performed on only one side. We have since performed bilateral TMR in one patient who presented with bilateral amputations caused by electrical burns (a left shoulder disarticulation amputation and a right transhumeral amputation).

The average duration between amputation and TMR surgery was 16 months, with a range of 4 months to 6 years. Under the initial experimental protocol, we only performed TMR on individuals with recent amputations (all of the first four patients had surgery within approximately 12 months of their amputation). This conservative approach was taken to ensure the optimal health of the donor nerves, as some axons in severed nerves can become nonviable over time (Kawamura and Dyck 1981). However, the robust success of early cases allayed this concern. Consequently, we are now willing to perform TMR on patients many years after injury; the TMR procedure that was performed 6 years after the original amputation resulted in good reinnervation of target muscle.

9.1.2 Acute Surgical Outcomes

Four nerve transfers were performed on all shoulder disarticulation patients, and two or three transfers were performed on transhumeral patients, depending on the amount of remaining brachialis muscle. Surgical procedures for shoulder disarticulation patients were of longer duration than those for transhumeral patients. The mean operative time for TMR procedures performed in shoulder disarticulation patients was 5 hours and 37 minutes, compared to 3 hours and 23 minutes for transhumeral cases. Increased operative times for shoulder disarticulation patients reflect the challenges imposed by altered anatomy due to injuries from high-impact trauma, which may result in the need to locate or create alternative muscle targets. These operative times also include additional surgical time for soft tissue coverage and other accessory procedures (such as dealing with heterotopic bone or extensive scarring), which are more frequently necessary in shoulder disarticulation patients.

Most of the patients remained in the hospital for one night following surgery. One patient was discharged the day of surgery. Three patients (10%) were kept longer—one for two nights, and two for three nights. One of these patients had poor acute pain control in a setting of chronic right leg pain. Another had an angulation osteotomy and soft tissue breakdown, which likely contributed to the extended hospitalization.

Delayed healing of the surgical incision was experienced by 3 out of the 27 patients; however, all of these patients went on to heal uneventfully without the need for additional revision procedures. Despite the extensive dissection and soft tissue mobilization required during TMR, none of these patients developed a clinically detectable hematoma, seroma, or infection following the procedure. This is likely due to the use of closed-suction drains and gentle external compression in the immediate postoperative period. Drains were generally removed once the patient had resumed normal activity and output had fallen below 30 ml/day. All patients received a standard 24-hour course of perioperative antibiotics, with intraoperative antibiotic irrigation of the wound prior to closure.

9.1.3 Long-Term Surgical Outcomes

Long-term surgical outcome data were available for all 27 patients who underwent TMR procedures at NMH and SAMMC. The mean duration of follow-up was 28 months. In all, a total of 79 nerve transfers were performed, yielding robust EMG signals in 75 target muscles—a 95% success rate. However, not all successful nerve transfers generated signals that could be used for prosthesis control, due to issues such as EMG signal cross talk from adjacent muscles or difficulty keeping an electrode securely placed over the target muscle.

Two of the unsuccessful nerve transfers involved transfer of the ulnar nerve to the pectoralis minor muscle. In at least one of these cases, failure was attributed to vascular compromise of pectoralis minor, which likely occurred during mobilization of the muscle segment into the axilla (Kuiken et al. 2004). The two other failed transfers occurred in a patient with a history suggestive of an avulsive mechanism of amputation. In this case, the radial nerve was found to be grossly abnormal (i.e., very small and atrophied) at the time of transfer. It is likely that this patient had a brachial plexopathy that had not been diagnosed preoperatively (O'Shaughnessy et al. 2008).

Patients were generally able to resume using their original prostheses 1 to 2 months after surgery—after the edema had resolved and after necessary modifications were made to socket fit and prosthesis control strategy. (See Chapter 6 for a discussion of prosthetic fitting after TMR.) In most cases, muscle contractions in the reinnervated muscle segments became sufficiently robust to allow for fitting with a TMR-controlled myoelectric prosthesis between 6 and 7 months after surgery. Another notable benefit of surgery was that the EMG signals from the natively innervated biceps and triceps muscle segments were often larger after TMR because subcutaneous fat over these muscles was removed during surgery. This facilitated subsequent prosthetic fittings.

Of the 27 patients who underwent TMR at NMH/RIC and SAMMC, 25 (93%) were subsequently fit with TMR-controlled myoelectric devices. The two patients who were not fit included the previously mentioned patient with brachial plexopathy and an additional patient who discontinued his rehabilitation and prosthetic fitting due to social reasons. Of note, the patient with the brachial plexopathy was able to wear and operate a conventional myoelectric prosthesis using EMG signals from his natively innervated biceps and triceps muscles.

9.1.4 Postsurgical Pain Outcomes

Outcomes related to neurogenic pain are of particular importance given the prevalence of neuroma and phantom limb pain in the TMR target population. Fifteen of the 19 patients who presented with phantom limb pain experienced an exacerbation of phantom limb pain symptoms following the procedure. The phantom limb pain increased significantly after surgery, but it did not reach the level experienced after the initial amputation. Since an increase in phantom limb pain was expected, it was immediately treated with neuropathic pain medications. In general, phantom limb pain returned to baseline levels within 4 to 6 weeks. One patient reported a significant increase in phantom limb pain that persisted for several months after surgery. In four patients, the quality and severity of their preoperative phantom limb pain remained relatively unchanged following TMR.

Fourteen patients presented with painful neuromas prior to surgery. Of those 14 patients, neuroma pain persisted in only 2 patients following TMR surgery. Interestingly, both patients with refractory neuroma pain underwent TMR for transhumeral-level amputations, and pain was generally found to be localized to the distal aspect of the residual limb. Both patients were diagnosed as having residual neuromas of the lateral antebrachial cutaneous nerve, a nerve not transferred as part of the conventional transhumeral TMR procedure. Although conclusions are limited by the small sample size and the retrospective nature of this review, these outcomes suggest a therapeutic role for TMR in the management of neuroma pain. (See Chapter 4 for a description of the possible role of TMR in neuroma prevention and treatment.)

9.2 Assessment of Functional Outcomes at RIC

The goal of TMR is to create additional physiologically appropriate myoelectric control sites to allow easier, more natural control of a prosthesis. Evaluating changes in prosthetic function after TMR posed a challenge when the technique was first developed due to the lack of validated functional outcome measures for adult upper limb amputees (Wright 2006). Three objective outcome measures—a modified Box and Block Test, a Clothespin Relocation Test, and the Assessment of Motor and Process Skills (AMPS); and one subjective outcome measure—the Disability of the Arm, Shoulder, and Hand Survey (see Box 9.1)—were selected on the basis of their expected ability to

BOX 9.1   OUTCOME MEASURES USED IN INITIAL
FUNCTIONAL TESTING OF TMR SUBJECTS

Objective Outcome Measures

The Box and Block Test (Mathiowetz et al. 1985) is a validated and standardized measure of gross hand function. The test consists of the subject moving 1-inch blocks from one side of a box to the other, over a low vertical partition, in a specified time frame. To assess function before and after TMR, the test was modified in two ways: (1) the amount of time allotted to the task was increased from 1 minute to 2 minutes so that more blocks could be moved and differences more easily observed, and (2) the task was performed while standing instead of sitting as individuals with a shoulder disarticulation amputation could not maneuver their prostheses effectively while sitting. This task was chosen because it was expected that after TMR, performance speed would be improved due to easier, simultaneous actuation of elbow and hand without the user having to switch between functions.

The Clothespin Relocation Test was adapted from the Roylan Graded Pinch Exerciser (Kuiken et al. 2004). Individuals were timed while moving three clothespins from a horizontal bar, rotating them, and placing them on a vertical bar. This task was chosen because it requires control of all three available degrees of freedom of the prosthesis, including wrist rotation. Again, it was expected that TMR would enable these individuals to complete this task more efficiently.

AMPS testing (Fisher 1993, 1994, 2003) can be used to assess effort, safety, efficiency, and independence in task performance. Individuals are asked to select and perform at least two familiar, relevant activities of daily living from a list of more than a hundred standardized activities. Subjects are scored according to their level of skill at moving themselves or an object (motor skills) or on their cognitive ability to plan the task, to interact with and use tools appropriately, and to solve problems (process skills). Raw scores are converted to linear measures of ability (logits) taking into account the predetermined relative "challenge" of the selected task and the relative stringency of the scoring by the clinician. A motor skill logit score below 2.0 indicates problems with the execution of the task; a process score below 1.0 indicates problems with the effectiveness of task performance. A difference of 0.5 logits is considered significant. This test was chosen because it was expected that TMR would decrease the cognitive load involved in operating the device, which would allow the user to plan tasks and pre-position the device more effectively. Note: AMPS testing is not specifically designed or validated to assess prosthesis use and cannot be used to assess function in bilateral amputees. Also, relative challenge may be related to planning requirements, number of steps, or choices but is not necessarily related to prosthesis function or control or obstacles that hinder use of a prosthetic device.

Subjective Outcome Measure

The Disabilities of the Arm, Shoulder, and Hand (DASH) outcome measure (Hudak et al. 1996; Beaton et al. 2001) is a self-report questionnaire that is designed to measure physical function and symptoms in individuals with any of several upper limb disorders—it is not specific to individuals who have had amputations. Subjects are questioned on their ability to perform tasks and the degree to which their disability has impacted social activities, work, or activities of daily living. Tasks can be performed with an assistive device, such as a prosthesis. Subjects answer 30 questions by selecting one of five answers with a corresponding score. Scores are averaged and converted to a DASH score; a higher score indicates a greater degree of disability. Optional, modular sections on work and sport are included.

FIGURE 9.1
Timed tasks used to evaluate function before and after TMR surgery. (a) Box and Block Test: subject moves blocks from one side of a vertical divide to the other; (b) Clothespin Relocation Test: subject moves clothespins from a horizontal bar to a vertical bar. (Reprinted from Miller et al., Improved myoelectric prosthesis control using targeted reinnervation surgery: a case series, *IEEE Trans Neural Syst Rehabil Eng* 16 (1):46–50, 2008. © [2008] IEEE. With permission.)

detect relevant changes in functional ability after TMR. (See Chapter 7 for currently recommended outcome measues.)

The Box and Block Test (Figure 9.1a) and Clothespin Relocation Test (Figure 9.1b) are simple physical performance metrics that provide concrete outcome parameters. It was expected that after TMR, users would be able to complete these tasks more quickly if TMR provided more intuitive control and allowed simultaneous control of both the hand and elbow, thus reducing or eliminating the need for mode switching. The Box and Block Test requires use of only the elbow and hand. The Clothespin Relocation Test is more difficult, as it requires use of wrist rotation to complete the task.

AMPS testing was used to assess the patient's performance of self-selected activities of daily living (ADLs). This test gives an indication of functional performance in real-world situations (see Box 9.1). However, AMPS testing is not specific to amputees; in fact, for the purposes of this test, a prosthesis is considered to be an assistive device, and the scoring method may actually discriminate against use of a prosthesis if the task can be completed more efficiently using one hand. However, pre-positioning of a multi-joint prosthesis for ADLs was considered likely to impose a considerable cognitive load. It was expected that if TMR provided easier, more intuitive control, this cognitive load would be reduced, allowing the user to focus more on pre-positioning and planning and thereby obtain higher process scores.

In addition to completing the objective tests, three of these initial subjects were asked to complete the DASH questionnaire, a more subjective outcome measure (see Box 9.1). Self-reports and observations by several of the TMR patients were also obtained through questionnaires before and after surgery, and patient comments during and after treatment were recorded. These more subjective measures provided additional data to assess ease of prosthesis use after TMR and how this affected the user's ability and desire to use the prosthesis for real-life tasks.

9.2.1 Functional Outcomes

Six individuals underwent functional testing at RIC before and after TMR (Miller et al. 2008). Three of these individuals had transhumeral amputations (T1, T2, and T3), two had shoulder disarticulation amputations (S1 and S3), and one patient (S2) had a very proximal transhumeral amputation with a very short (3-cm) residual humerus. S2 was thus treated as a shoulder disarticulation amputee for the purposes of TMR surgery and prosthetic fitting.

All subjects except S2 were fitted at RIC prior to surgery. This fitting included a Boston Digital Arm (Liberating Technologies, Inc.), an electric wrist rotator (Otto Bock), and an electric terminal device (Otto Bock). Subject S2 was fitted elsewhere before surgery with a Utah arm (Motion Control Inc.) and was fitted again after surgery with a new device comprising the same components listed above.

Pre-TMR control strategies varied widely for the shoulder disarticulation subjects (Table 9.1). The transhumeral subjects used similar control strategies before TMR: T1 and T2 used two-site EMG to control elbow and hand and a linear transducer to control the wrist; T3 used EMG control for all components. Where necessary, all subjects achieved mode selection by co-contraction, time-out, or bump-switch strategies. Subjects received occupational

TABLE 9.1

Pre-TMR Control Strategies for Shoulder Disarticulation Patients

Subject	Elbow Flexion/Extension	Wrist Rotation	Hand Open/Close
S1	Socket-mounted FSR (superior) fast/slow	Socket-mounted FSR (posterior) fast/slow	Two socket-mounted FSRs (both anterior)
S2[a]	Two-site EMG: p. major and triceps	passive	Two-site EMG: p. major and triceps
S3	Anterior FSR: Elbow flexion, wrist supination, hand close Posterior FSR: Elbow extension, wrist pronation, hand open		

Note: EMG, electromyographic signal; FSR, force-sensing resistor.
[a] Subject was fitted with a Utah Arm (Motion Control Inc.).

therapy training to ensure that the devices functioned properly and that they were able to operate the devices successfully. After training and at least 8 months of practice at home, subjects performed the Box and Block Test, Clothespin Relocation Test, and AMPS assessment. (Note: S1 had bilateral shoulder disarticulation amputations and so was unable to perform AMPS testing. S2 was unable to perform the Clothespin Relocation Test before TMR because her pre-TMR device had a passive wrist.)

Post-TMR control strategies for shoulder disarticulation subjects are shown in Table 9.2. Nerve transfers yielded four control signals in each of these subjects. In the transhumeral subjects, four control sites for elbow flexion/extension and hand open/close were obtained through a combination of nerve transfers and native innervation. The median nerve was transferred to the short head of the biceps and the distal radial nerve was transferred to either the brachialis muscle (T1 and T2) (O'Shaughnessy et al. 2008) or to the lateral head of the triceps (T3) (Dumanian et al. 2009). The natively innervated long head of the biceps and long and medial heads of the triceps provided signals for elbow flexion and elbow extension, respectively. Wrist rotation was controlled by a harness-mounted linear potentiometer.

All subjects used the same device after reinnervation with minor modifications to socket fit. Changes were made to the control strategy to accommodate the additional control sites created by TMR and the additional electrodes. Appropriate software modifications were made to accomodate these additional inputs and the functions they controlled. Thus, with the exception of S2, functional control of the same device could be directly compared before and after TMR. In the case of S2, changes in function due to differences in devices, in addition to TMR, cannot be ruled out.

Subjects received occupational therapy training to use the TMR-controlled device. (See Chapter 7 for a recommended training protocol for TMR patients.) Less training was needed in order to establish good control of the device using sites created by TMR, likely reflecting the more intuitive nature of this control. Once TMR-based control of the device was mastered, functional testing was performed after approximately 1, 3, and 6 months of home use.

TABLE 9.2

Nerve Transfers and Target Muscles Used to Control Prosthetic Function after TMR in Shoulder Disarticulation Subjects

Subject	Elbow Flexion	Elbow Extension	Hand Close	Hand Open	Wrist Rotation
S1	Musculocutaneous nerve to clavicular head of p. major	Radial nerve to lower sternal head of p. major	Median nerve to upper sternal head of p. major	Median nerve to upper sternal head of p. major[a]	Switch between hand and wrist control using a socket-mounted FSR
S2[b]	Musculocutaneous nerve to clavicular head of p. major	Native innervation of remnant triceps	Median nerve to sternal head of p. major and p. minor[c]	Radial nerve to thoracodorsal nerve[d]	Two-site control using two socket-mounted FSRs
S3	Musculocutaneous nerve to clavicular head of p. major	Native innervation of triceps/posterior deltoid	Ulnar nerve to sternal head of p. major	Radial nerve to p. minor	Single-site control using one socket-mounted FSR

Note: Reprinted and adapted from Miller et al., Improved myoelectric prosthesis control using targeted reinnervation surgery: a case series, *IEEE Trans Neural Syst Rehabil Eng* 16 (1):46–50, 2008. © [2008] IEEE. With permission.

[a] The median nerve transfer fortuitously gave rise to two independent signals—attempted thumb abduction was used to provide a signal for hand open.

[b] Humeral neck amputation treated as a shoulder disarticulation amputation.

[c] Median nerve was split longitudinally and transferred to two separate target muscles in an attempt to generate two independent signals. This was not successful; did not yield two independent signals.

[d] Target muscle: serratus anterior.

After approximately 1 month of using the TMR-controlled prosthesis at home, subjects were able to move an average of 323% (SD 151%) more blocks in the allotted time than when using the conventional device, even after considerably longer experience with that device (Figure 9.2a). Subjects were also able to manipulate and relocate clothespins an average of 49% (SD 12%) more quickly than they had been able to do using conventional control (Figure 9.2b).

AMPS testing was performed to assess the subject's ability to independently perform two self-selected tasks. Raw scores were converted to logit scores—linear measures of ability that take into account task complexity and variability of clinician scoring. Motor scores reflect the subject's level of skill at moving objects and moving themselves; a score below 2.0 indicates difficulty in this area. Process scores reflect the subject's ability to plan the task, solve problems, and use tools, and were expected to reflect the cognitive load imposed by use of the prosthesis; a score below 1.0 indicates inadequate performance.

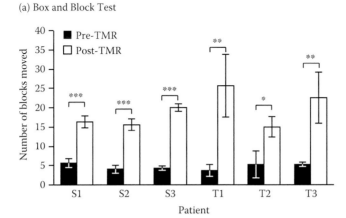

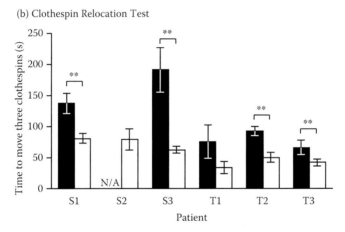

FIGURE 9.2

(a) Average number of blocks transferred during three trials using either a conventionally controlled prosthesis or a TMR-controlled prosthesis. (b) Time taken by subjects to move three clothespins from a horizontal bar to a vertical bar: average of three trials using a conventionally controlled device or a TMR-controlled device. Note: Subject S2 was unable to perform the clothespin relocation test prior to undergoing TMR because her preoperative device did not have a wrist rotator. Subjects had used conventional control for various periods (at least 8 months) prior to testing. TMR control was tested after approximately 1 month of use. (Reprinted and adapted from Miller et al., Improved myoelectric prosthesis control using targeted reinnervation surgery: a case series, *IEEE Trans Neural Syst Rehabil Eng* 16 (1):46–50, 2008. © [2008] IEEE. With permission.)

Most subjects showed increased logit scores in both motor and process skills after TMR (Figure 9.3). Four out of five subjects demonstrated significant improvement (i.e., an increase of greater than 0.5 logits) in motor scores, although none achieved a score of 2.0, indicating that all had motor skill difficulty when using any prosthesis (Figure 9.3a). Three of the five subjects

(a) Motor scores

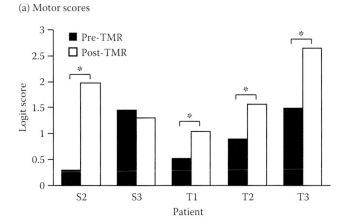

(b) Process scores

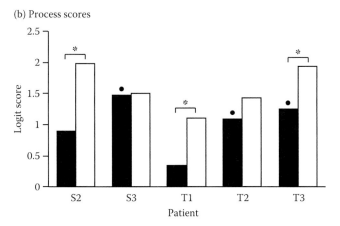

FIGURE 9.3
Logit scores from AMPS testing of TMR subjects: (a) Motor scores, (b) Process scores. * indicates significant improvement, • indicates subjects with pre-TMR process scores greater than 1.0. Subjects had used conventional control for various periods (at least 8 months) prior to testing. TMR control was tested after approximately 1 month of use. SD01 is a bilateral amputee and so could not undergo AMPS testing. (Reprinted and adapted from Miller et al., Improved myoelectric prosthesis control using targeted reinnervation surgery: a case series, *IEEE Trans Neural Syst Rehabil Eng* 16 (1):46–50, 2008. © [2008] IEEE. With permission.)

had process scores before TMR that did not show detectable cognitive deficits (i.e., initial process scores using the non-TMR prosthesis were above the cutoff score of 1.0). Thus process scores did not adequately reflect any cognitive burden before TMR in these subjects and, due to this ceiling effect, any change after TMR could not be quantified. Two subjects who had initial process scores below 1.0 (S2 and T1) did show significant improvement after TMR and achieved process scores greater than 1.0 (Figure 9.3b).

TABLE 9.3

DASH Scores Obtained before and after TMR

Subject	Pre-TMR	Post-TMR
T3	41.6	39.16
S2	23.3	20
S3	29.16	22

9.2.2 Patient-Reported Outcomes

Three subjects were evaluated using the DASH outcome measure. The DASH consists of 30 scored questions on physical function, symptoms, and social functioning.* A higher score indicates a higher level of disability. All subjects tested showed improvement when using the TMR-controlled prosthesis (Table 9.3).

Several subjects reported that tasks—including yard work, home maintenance, and cooking—were easier to perform, and that some new tasks were now possible. In addition, some subjects reported that they used their prosthesis more frequently and for longer periods. Subjects were asked to provide subjective opinions on the effects of TMR. Selected opinions are provided below:

- S1 reported a "night and day" difference between his two devices, finding the TMR-controlled prosthesis easier and faster to use and requiring less conscious thought. "You don't have to think about it, it's there. I move my phantom limb—that's how I operate it. I move my phantom limb and it responds"; "I guess in my mind, my hand is still there, when I open that, I'm literally...opened [*sic*] my hand. And when I close it, I literally close my hand." Referring to his previous device, S1 reported, "It was aggravating, it was hard to use, I used the hook (body-powered side) most of the time." "This is the future right here. It's not perfect, but it sure beats what I had."

  S1 felt that the operation of the prosthesis was noticeably smoother after TMR: "I can pick up objects...like a quarter. I can pick it up flat off the floor or a table.... I can pick up an object like a round ball or something like that. I can actually snatch it up in a short time." "[It] responds quickly; if there's a delay I don't notice it." Referring to flexing his elbow, he commented, "Wherever I want to stop it, 99.9% of the time it stops where I want it." This subject reported that several tasks were either easier to perform, or became possible to perform, with the TMR-controlled device (Kuiken et al. 2004).

---

* Disabilities of Arm, Shoulder and Hand Outcome Measure. The Institute for Work and Health. Available at http://www.dash.iwh.on.ca/home, Accessed October 25, 2012.

- S2 reported a marked improvement in her ability to control her prosthesis after TMR. She reported wearing the TMR-controlled device an average of 4 to 5 hours per day (up to 16 hours at a time), 5 to 6 days per week (Kuiken et al. 2007).

- S3 commented, "It changes the way I do things because it is more efficient. It is easier to control than the previous prosthesis." "This is much easier, more maneuverable. [It's] faster, easier to use."

- T1 and T2 reported that they preferred the function of the TMR-controlled prosthesis to that of their previous devices and indicated that operation of the new device was smoother and felt more natural (O'Shaughnessy et al. 2008). Both individuals reported using their TMR-controlled devices regularly. T1 used his TMR-controlled prosthesis for 3 to 10 hours per day, 3 to 5 days a week, and was able to return to work. However, he stated that both pre- and post-TMR prostheses were heavy and that the weight and difficulty of donning the device deterred him from wearing it more frequently. T2 reported using the device for 3 to 6 hours per day, 4 to 5 days a week. T2 commented, "It doesn't change the way I do things, but it is faster and this prosthesis is easier to control."

- Before TMR, T3 used EMG signals from her biceps and triceps muscles to control the prosthetic elbow and hand. Mode selection was achieved through co-contraction. T3 reported that it "wasn't very normal." T3 also commented, "[The] other [pre-TMR] prosthesis was easy to work but switching was awkward and slow. This is easy, so yes, it changes the way I do things." She also stated, "Moving elbow and hand—it's become a very natural thing. This prosthesis is more effective and easier to control." "To be able to think 'open your hand' and your hand opens, rather than using your elbow, which is so unnatural—I wouldn't change it for anything." T3 reported using the TMR-controlled device primarily at home: "I use it when I need to—to carry things around...things I can't do with one hand... laundry, cooking...I wear it around the house...but it's heavy." T3 indicated that she did not use the device socially, for example, when going to a restaurant, and that more than 2 or 3 hours of use caused shoulder pain.

9.3 Conclusion

Since 2002, TMR has been performed on more than 60 individuals in more than a dozen institutions throughout the world. Our experience suggests that TMR is a safe, effective, and well-tolerated procedure with minimal surgical morbidity. In light of the potential for enhanced prosthetic control that TMR offers, and the additional possibility of reduced neuroma pain, the relatively minor risks of this surgical procedure should not serve as a barrier to widespread application of the technique.

Despite the limited availability of suitable outcome measures to assess functional improvement after TMR, both objective testing and subjects' subjective opinions indicate that TMR provides faster, more intuitive prosthesis control for high-level upper limb amputees. Longer-term follow-up is required to determine whether the benefits provided by TMR translate into increased prosthesis use over time. TMR only improves the control of the prosthesis; it does not address other common issues that result in low prosthesis usage, such as prosthesis weight and comfort. Our hope is that TMR will provide the impetus for the development of lighter, more functional prosthetic devices, which will allow users to more fully realize the benefits of improved control.

References

Beaton, D. E., J. N. Katz, A. H. Fossel, J. G. Wright, V. Tarasuk, and C. Bombardier. 2001. Measuring the whole or the parts? Validity, reliability, and responsiveness of the Disabilities of the Arm, Shoulder and Hand outcome measure in different regions of the upper extremity. *J of Hand Ther* 14 (2):128–146.

Dumanian, G. A., J. H. Ko, K. D. O'Shaughnessy, P. S. Kim, C. J. Wilson, and T. A. Kuiken. 2009. Targeted reinnervation for transhumeral amputees: current surgical technique and update on results. *Plast Reconstr Surg* 124 (3):863–869.

Fisher, A. G. 1993. The assessment of IADL motor skills: an application of many-faceted Rasch analysis. *Am J Occup Ther* 47:319–329.

Fisher, A. G. 1994. Development of a functional assessment that adjusts ability measures for task simplicity and rater leniency. In *Objective Measurement: Theory into Practice*, edited by M. Wilson. Norwood, NJ: Ablex.

Fisher, A. G. 2003. *AMPS Assessment of Motor and Process Skills*, 5th ed. Fort Collins, CO: Three Star Press.

Hudak, P. L., P. C. Amadio, C. Bombardier, et al. 1996. Development of an upper extremity outcome measure: the DASH (Disabilities of the Arm, Shoulder, and Head). *Am J Ind Med* 29 (6):602–608.

Kawamura, Y., and P. J. Dyck. 1981. Permanent axotomy by amputation results in loss of motor neurons in man. *J Neuropathol Exp Neurol* 40 (6):658–666.

Kim, P. S., J. H. Ko, K. D. O'Shaughnessy, T. A. Kuiken, E. A. Pohlmeyer, and G. A. Dumanian. 2012. The effects of targeted muscle reinnervation on neuromas in a rabbit rectus abdominis flap model. *J Hand Surg* 37 (8):1609–1616.

Kuiken, T. A., G. A. Dumanian, R. D. Lipschutz, L. A. Miller, and K. A. Stubblefield. 2004. The use of targeted muscle reinnervation for improved myoelectric prosthesis control in a bilateral shoulder disarticulation amputee. *Prosthet Orthot Int* 28 (3):245–253.

Kuiken, T. A., L. A. Miller, R. D. Lipschutz, et al. 2007. Targeted reinnervation for enhanced prosthetic arm function in a woman with a proximal amputation: a case study. *Lancet* 369 (9559):371–380.

Mathiowetz, V., G. Volland, N. Kashman, and K. Weber. 1985. Adult norms for the Box and Block Test of manual dexterity. *Am J Occup Ther* 39 (6):386–391.

Miller, L. A., K. A. Stubblefield, R. D. Lipschutz, B. A. Lock, and T. A. Kuiken. 2008. Improved myoelectric prosthesis control using targeted reinnervation surgery: a case series. *IEEE Trans Neural Syst Rehabil Eng* 16 (1):46–50.

O'Shaughnessy, K. D., G. A. Dumanian, R. D. Lipschutz, L. A. Miller, K. Stubblefield, and T. A. Kuiken. 2008. Targeted reinnervation to improve prosthesis control in transhumeral amputees. A report of three cases. *J Bone Joint Surg Am* 90 (2):393–400.

Wright, V. F. 2006. Measurement of functional outcome with individuals who use upper extremity prosthetic devices: current and future directions. *J Prosthet Orthot* 18 (2):46–56.

10

Future Research Directions

Levi J. Hargrove and Blair A. Lock

CONTENTS

10.0 Introduction

Targeted muscle reinnervation (TMR) has rapidly transitioned from an experimental procedure—first performed on a human patient in 2002 (Kuiken et al. 2004)—to accepted clinical practice. It is now performed worldwide on individuals with transhumeral and shoulder disarticulation amputations. When coupled with the rehabilitation strategies described in preceding chapters, TMR provides significant functional improvements and increased quality of life for individuals with high-level upper limb amputations. However, current functional control has many limitations, and the individuals currently served by TMR represent only a small proportion of the total population liv-ing with limb loss. There is a tremendous opportunity to improve the control paradigms used for TMR prostheses and to extend the TMR surgical tech-nique to individuals with below-elbow and lower limb amputations. TMR has also enabled advanced research and development opportunities, includ-ing development of advanced electromechanical prostheses, novel training routines, and innovative fitting protocols. In combination, these advances could result in an unprecedented ability to control a prosthesis for the major-ity of individuals with major amputations.

10.1 Improvements in Control Algorithms

10.1.1 Limitations of Current Control Algorithms

Although current electromyographic (EMG) control techniques provide rudimentary control of powered prostheses, their function is highly limited. The rather primitive algorithms traditionally used for myoelectric control are based only on the magnitude of the EMG signals; as a result, they impose restrictions on the placement of recording electrodes and extract only a lim-ited amount of the available neural information. (See Williams 2004 for a summary of conventional myoelectric control strategies.)

Surface EMG signals provide a global picture of muscle activity and often comprise signals from many muscles (known as *muscle cross talk*) that result from co-activation of different muscles in order to perform certain tasks. Muscle cross talk confounds conventional EMG-amplitude-based control algorithms; for robust operation, conventional control techniques require electrodes to be placed over independent muscle groups that are free from cross talk. These systems work best when (1) electrodes are placed over ago-nist/antagonist muscle pairs and (2) the resulting signals are used to control a physiologically appropriate prosthesis movement. For example, electrodes placed over the biceps and triceps muscles are ideally suited to control elbow flexion/extension but cannot easily be used for hand or wrist con-trol. Unfortunately, above-elbow amputation leaves very few suitable con-trol sites. Even after TMR surgery, locating four independent sites for control of two prosthetic joint movements is challenging and requires a significant amount of trial and error.

TMR involves the transfer of large mixed nerves that are capable of trans-mitting an enormous amount of motor control information from the brain to muscles. As a result, reinnervated muscles contain a great deal of neural con-trol information. Theoretically, the neural activation patterns of muscles after TMR should be diverse and contain contributions from motor neurons that previously innervated all muscles of the arm, including the intrinsic muscles

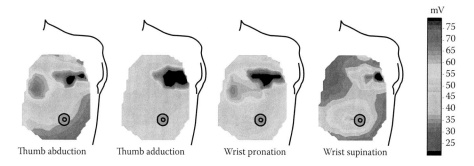

Thumb abduction | Thumb adduction | Wrist pronation | Wrist supination

FIGURE 10.1
(See color insert.) Mean absolute values of the EMG signal amplitudes from 115 high-density electrodes generated when a TMR patient attempted the indicated test contractions.

of the hand. The diversity of the neural information transmitted to reinnervated muscles has been confirmed visually and through palpation of the reinnervated sites during the performance of test contractions (e.g., flexing the elbow or closing the hand) by TMR patients. High-density electromyography, in which the reinnervated musculature is saturated with large numbers of electrodes, is a quantitative method of exploring the neural content of reinnervated muscles. Visual representations of the resulting muscle activation patterns have been created by plotting the amplitudes of the underlying signals as the TMR patient makes different contractions (Figure 10.1). These studies have confirmed that after TMR, reinnervated muscles contain a plethora of neural control information, including that corresponding to thumb, fingers, and wrist movement (Zhou et al. 2007). Unfortunately, conventional control algorithms are unable to decode this information.

10.1.2 Overview of Pattern Recognition Myoelectric Control

Pattern recognition algorithms can be used to decode EMG signal patterns to determine an individual's intended movements. These algorithms have been investigated for decades and are conceptually straightforward. (See Scheme and Englehart 2011 for detailed descriptions of each component of the pattern recognition algorithm.) For example, an array of electrodes placed around the arm of a transhumeral amputee with TMR is adequate for robust pattern recognition control; the user can control elbow flexion/extension, wrist flexion/extension, wrist rotation, and even different hand grasps. The operation of these functions is intuitive because the pattern recognition system recognizes what the user wants to do and then sends the appropriate control signals to the prosthesis.

Although there are a variety of pattern recognition algorithms, they all operate using the same sequence of steps (Figure 10.2). Using machine-learning techniques, a computer program first "learns" an individual's EMG signal patterns for a set of intended movements by recording examples of

FIGURE 10.2
Diagram of the major steps in the EMG pattern recognition process.

EMG signals generated for each of those movements; these EMG signals are often called *training data*. As the individual performs this predefined series of attempted movements, EMG signals are recorded and divided into short segments of 50 to 250 ms (called *data windows*). Distinctive elements of the signal (called *features*) are then extracted from each data window. EMG signal features are generally related to the amplitude or frequency of the signal, such as the maximum absolute magnitude, the number of times the EMG signal changes direction (called *turns*), or how often the signal crosses the zero baseline (called *zero crossings*). These features are quantified for each data window and then sent to a mathematical algorithm that sorts through the feature sets from hundreds of data windows for each EMG channel and determines an appropriate mathematical model for classifying the features sets according to the corresponding intended movement. This mathematical model is known as the *classifier*. Once this classifier is trained (i.e., a custom classification system is built for that user), it can be used for prosthesis control. Subsequent EMG patterns are again divided into data windows, the features are quantified, the classifier decodes user intent, and the desired movement is initiated.

The performance of pattern recognition systems is most often quantified in terms of *classification error*—the percentage of time the system incorrectly predicts the intended movement. Pattern recognition systems do not require electrodes to be placed exactly over independent agonist/antagonist muscle pairs. In fact, they work best with signals from many muscles, and cross talk is beneficial as opposed to problematic (as it is for conventional control). Pattern recognition allows for control of more prosthetic movements than conventional amplitude-based control techniques. Many common data features and classifiers have been investigated and have been found to perform very well for pattern recognition (Hargrove et al. 2007).

10.1.3 Performance of Pattern Recognition Systems after TMR

The EMG signal patterns for individuals with TMR have been examined in detail (Zhou et al. 2007). High-density EMG signals were collected using between 79 and 128 electrodes from four individuals as they attempted to complete 16 different arm, wrist, and finger movements. The data were examined post hoc. On average, the pattern recognition system was able to recognize the movements with less than 5% classification error. However, it is not practical to construct sockets containing such high numbers of electrodes.

Further experiments demonstrated that excellent classification performance could be achieved using only 4 to 12 pairs of closely spaced (less than ~2 cm) electrodes placed strategically over reinnervated muscles (Huang et al. 2008). If electrode pairs with wider spacing (~3.5 cm) are used, then a grid of electrodes over the reinnervated muscles is sufficient, and strategic placement is unnecessary (Tkach et al. 2012).

10.1.3.1 Pattern Recognition in Virtual Environments

The EMG experiments described in the previous paragraph demonstrated that individuals can reliably generate consistent EMG signal patterns after TMR. Furthermore, we have shown that these signal patterns can be recognized using a clinically viable number of electrode pairs; however, it is important to evaluate the individual's ability to use these patterns to control a prosthesis in the presence of real-time feedback. Virtual environments— interactive computer programs that allow subjects to control a prosthesis on a computer screen—have proven to be excellent tools for training and testing pattern recognition control. A virtual prosthesis can be used without the added complexity and weight of a physical prosthesis. A complex virtual prosthesis with multifunctional hands and wrists can be used even if a physical version does not exist or is unreliable. Furthermore, many parameters can be easily changed in the virtual environment. Virtual environment performance tests are also useful tools for evaluating an individual's ability to control a device, as movements can be more easily quantified in the virtual environment. Two real-time virtual performance tests have been used recently to measure EMG pattern recognition control by TMR patients: the Motion Test (Kuiken et al. 2009) and the Target Achievement Control (TAC) Test (Simon et al. 2011b).

The Motion Test evaluates the individual's ability to move through the complete range of motion of a given movement. Results are quantified by measuring the time required to select a particular movement, the time taken to move through the entire range of motion of the movement, and the percentage of movements successfully achieved. After TMR, individuals using pattern recognition demonstrate performance that is comparable to that of age-matched non-amputee control subjects (Figure 10.3) (Kuiken et al. 2009). This provides further evidence that TMR coupled with pattern recognition provides excellent control.

The TAC Test evaluates an individual's ability to bring a virtual prosthesis into a target posture and maintain that posture for a specified period of time. The TAC Test can be made much more challenging than the Motion Test because the individual is required to coordinate different movements to achieve the target posture. For instance, one test might simply require the patient to flex the wrist by 40 degrees and remain within the targeted location (Figure 10.4). A more challenging multiple-degree-of-freedom test might require the user to flex the wrist to 30 degrees, pronate the forearm by 10

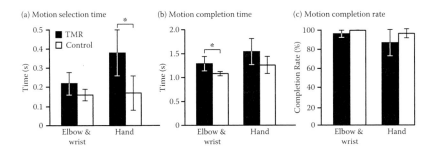

FIGURE 10.3
Results from a Motion Test with a 5-s time limit performed by transhumeral and shoulder dis-articulation amputees after TMR ($n = 5$) and non-amputee controls ($n = 5$). Error bars represent ± 1 SD. * indicates a significant difference ($p < 0.05$).

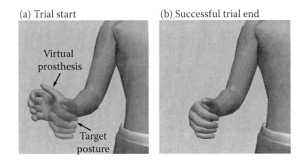

FIGURE 10.4
(See color insert.) A simple example of the TAC Test display presented to a user. The patient has real-time control over the virtual prosthesis and is required to move it to the target posture. In this example, the user is only required to perform wrist flexion to successfully complete the trial. (From Simon et al., Target Achievement Control Test: evaluating real-time myoelec-tric pattern-recognition control of multifunctional upper-limb prostheses, *J Rehabil Res Dev* 18 (6):619–628, 2011. With permission.)

degrees, and close the hand halfway (each with reasonable tolerance). The difficulty is compounded by the fact that the virtual environment does not allow the individual to use any compensatory body movements to help him or her reach the goal. The results of this test are quantified in terms of the time taken to reach the target and the percentage of successfully completed trials.

10.1.3.2 Pattern Recognition with Physical Prostheses

Pattern recognition control with physical prostheses was first demonstrated by a TMR patient in 2009 (Kuiken et al. 2009). Initially, the program was run on a desktop computer connected to the prosthesis by a cable. Later, the cable was replaced with a telemetry system. More recently, microcontrollers have been built into the prosthesis, enabling pattern recognition control in patients' home environments. The application of pattern recognition has also evolved in many ways. Initially, classifier training required a desktop computer and a virtual environment. Now, the prosthesis itself is used to guide training: patients follow the dynamic movements of the arm, and training is completed in less than 2 minutes (Simon et al. 2012). Other important refinements in velocity control and EMG signal collection have made the application of pattern recognition control faster and easier to learn and have provided more robust operation (Simon et al. 2011a; Young et al. 2012a).

Pattern recognition control of multiple degrees of freedom is much more intuitive than sequencing through functions using switches or co-contraction, as is necessary with conventional control. Unfortunately, it is difficult to develop quantitative performance tests that are representative of the diverse set of tasks completed during daily living. Consequently, simple timed performance tests that require the individual to interact with objects and make use of all prosthetic movements have been developed. In addition to the Box and Block Test and the Clothespin Relocation Test (see Chapter 9 for a description of these tests), a block-stacking test has been used in which individuals pick up and stack 1-inch cubes into as high a tower as possible within a given time frame. The block-stacking test requires individuals to demonstrate fine motor control by picking up blocks and placing them on the tower without knocking it over. It also requires that they use all available prosthetic movements: elbow flexion/extension, wrist rotation, wrist flexion/extension, and hand open/close. These three performance tests have been used to quantify improvements in TMR patients' control as a result of pattern recognition (Figure 10.5).

Qualitatively, all of the TMR patients who were tested strongly preferred using the pattern recognition system. They found it to be more intuitive to control, smoother to operate, and more consistent in its performance. Perhaps

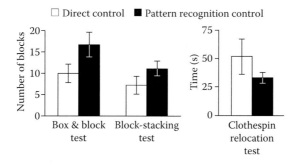

FIGURE 10.5
Average performance measures for TMR patients ($n = 3$) using conventional amplitude control (direct control) and pattern recognition control. Higher scores on the Box and Block Test and block-stacking test indicate better control. Lower scores on the Clothespin Relocation Test indicate better control. Error bars represent ± 1 SD.

most important, individuals using pattern recognition often requested that the technology be incorporated into their home-use prosthesis. This provides further motivation for the development of a commercially available pattern recognition solution.

10.1.3.3 Extension of Pattern Recognition to Simultaneous Control

One limitation of pattern recognition is that it can provide only sequential control of each prosthetic movement. Individuals must therefore plan the steps needed to execute each task. It is not straightforward to extend pattern recognition to include simultaneous control because a simultaneous movement is generally not a combination of individual discrete movements. However, if examples of EMG signals from combined motions are recorded, then pattern recognition algorithms are capable of learning these complex patterns (Young et al. 2012b). Qualitatively, simultaneous movements look more fluid and lifelike. The importance of restoring simultaneous control is shown quantitatively in Figure 10.6. Four TMR patients completed virtual environment TAC Tests that required them to move their limbs through two degrees of freedom to match a target posture. Conventional control allows simultaneous movements and outperforms pattern recognition, which is limited to sequential movements. However, using a pattern recognition system configured to recognize simultaneous

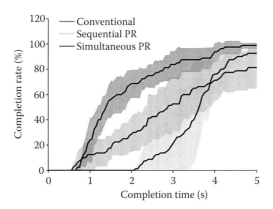

FIGURE 10.6
(See color insert.) Functional performance results averaged across four TMR patients as they completed TAC tests that required simultaneous control of two degrees of freedom. Each subject completed the tests using conventional TMR control (Conventional), sequential pattern recognition control (Sequential PR), and simultaneous pattern recognition control (Simultaneous PR). Higher and faster completion rates are representative of better control. Shaded regions represent ± 1 SD. (From Young et al., A functional comparison of simultaneous pattern recognition to conventional myoelectric control, *J Neural Eng Rehabil*. Under review.)

movements provides even better results. Simultaneous pattern recognition control remains an active area of research and it is likely that more controllable, multi-degree-of-freedom systems will be clinically available in the near future.

10.2 Extension of TMR to Transradial and Lower Limb Amputations

TMR has been used primarily for individuals with high-level upper limb amputation because of the significant functional deficits experienced by this population. However, in the near future, the benefits of TMR surgery may be extended to individuals with transradial, transfemoral, and transtibial amputations. This extension would dramatically increase the number of individuals that can benefit from the procedure.

10.2.1 TMR for Transradial Amputation

Between 1988 and 1996, there were 5,839 recorded transradial amputations in the United States, accounting for nearly 40% of all major (wrist and above) upper limb amputations and roughly equaling the combined number of shoulder disarticulation and transhumeral amputations (Dillingham et al. 2002). Transradial amputation often leaves significant residual limb musculature but, at most, only two suitable control sites to provide conventional amplitude-based myoelectric control. Even without TMR surgery, these individuals are ideal candidates for pattern recognition control, because many of the wrist-control and extrinsic hand- and finger-control muscles are still present. Transradial amputees without TMR have control of wrist movement that is similar to that of high-level upper limb amputees with TMR; however, they are not able to reliably perform as many hand-grasp patterns (Kuiken et al. 2009; Li et al. 2009). Figure 10.7 shows motion completion rates for transradial amputees performing five hand grasp patterns and for TMR subjects performing three hand grasp patterns. Even though different numbers of hand grasps were tested for each group, subsequent analysis of these data indicate that when three hand grasps are considered for each group, TMR subjects continue to demonstrate higher motion completion rates than transradial amputees. It is likely that the EMG signal patterns generated after TMR include contributions from nerves that controlled the intrinsic hand, finger, and thumb muscles, which makes different hand-grasp patterns easier to discriminate. These results are especially important because new multifunction hands that support multiple grasps are now commercially available. Studies are currently under way to quantify the information

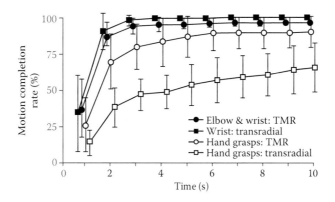

FIGURE 10.7
Real-time control achieved by individuals with TMR for shoulder disarticulation or trans-humeral amputation (*n* = 5) compared to individuals with a transradial amputation and no TMR (*n* = 5). Note: Individuals with TMR attempted three hand grasps; individuals with a transradial amputation attempted five hand grasps. Error bars represent ± 1 SD.

content provided by intrinsic hand muscles in order to guide future transradial TMR procedures.

10.2.2 TMR for Lower Limb Amputation

Lower limb amputation is increasingly common in the United States, leading to a rise in the demand for advanced prosthetic legs. It was estimated that 623,000 Americans were living with amputations of the foot or leg in 2005, accounting for almost two-thirds of all persons with amputations (Ziegler-Graham et al. 2008). Between 1988 and 1996, approximately 95% of transfemoral and transtibial amputations were caused by dysvascular disease, with the rate of this type of amputation increasing by 27% per year (Dillingham et al. 2002). Microprocessor-controlled powered upper limb prostheses have existed for many years, but microprocessor-controlled powered leg prostheses have only recently become commercially available (Au and Herr 2008; ÖSSUR). Various research groups are currently developing several others (Sup et al. 2008). Control of this type of device relies on the use of either mechanical sensors placed on the prosthesis, an external remote control, or an instrumented orthotic on the intact leg. Safe operation of these devices is critical, as patients may be seriously injured by a fall.

Recent studies have evaluated the suitability of EMG signals from residual leg muscles as inputs for control of powered leg prostheses. These studies have been completed using non-TMR amputees, passive devices (Huang et al. 2009; Huang et al. 2011), and virtual environments (Hargrove et al. 2011). The results of these studies have shown that EMG signals from residual thigh muscles contain useful control information. Furthermore, the neural information decoded from these signals can be combined with data from mechanical sensors embedded in the prosthesis. The resulting system with fused neural and mechanical information is more accurate and responsive in classifying different ambulation modes, such as level-ground walking or stair navigation.

There is great interest in using TMR to restore additional neural information to supplement available EMG signals and improve control of powered leg prostheses. Anatomic studies have shown that there are many muscles and motor points in the leg that would be suitable for use during TMR surgery (Agnew et al. 2012). Although TMR for advanced prosthesis control is not yet a clinical option for lower limb amputees, the procedure has been used for the prevention and treatment of painful neuromas in this patient population (Hargrove et al. 2013). (See Chapter 4 for discussion of the possible role of TMR in neuroma prevention and treatment.) One knee disarticulation amputee (TMR 1) and one transfemoral amputee (TMR 2), both of whom had TMR for clinical treatment of neuroma pain, participated in experiments to evaluate their ability to control powered lower limb prostheses during non-weight-bearing activities. High-density EMG experiments were completed to evaluate the potential of these individuals to control sagittal-plane movements of the knee and ankle, as well as femoral and tibial rotation. As expected, EMG signals were detected over the reinnervated muscles during test contractions. Both subjects achieved average classification errors of less than 5% when controlling the four movements: knee extension, knee flexion, dorsiflexion, and plantar flexion (Hargrove et al. 2013).

The individuals then completed real-time control experiments within a virtual environment using 11 bipolar electrodes placed over two reinnervated muscles and nine major muscles of the residual limb. Four transfemoral amputees who had not had TMR also completed the same experiment so that a baseline comparison could be made (Hargrove et al. 2011). Both TMR and non-TMR participants demonstrated excellent real-time control (Figure 10.8a). The ability to control a motorized knee and ankle during such circumstances would facilitate activities of daily living such as dressing or getting in and out of a vehicle. TMR 1 and the transfemoral amputees without TMR all demonstrated excellent real-time control of a system configured to recognize femoral rotation and tibial rotation in addition to the movements previously described (Figure 10.8b).

Motorized knees and ankles can provide functional benefits for tasks that are otherwise impossible with passive devices, such as walking up stairs with a reciprocating gait. To evaluate the control improvements provided by neural information, an experiment was completed in which TMR 1 completed 20 ambulation circuits around a gait laboratory. Each circuit was comprised of standing, level-ground walking, ramp ascent, ramp descent, stair ascent, and stair descent. The subject then completed 10 additional circuits while controlling the prosthesis in real-time using a pattern recognition system that interpreted only mechanical sensors or using a pattern recognition system that used both EMG signals and mechanical

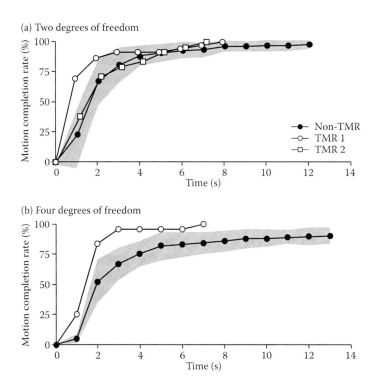

FIGURE 10.8
Real-time control results achieved during non-weight-bearing experiments to assess the control of (a) two or (b) four degrees of freedom by individual TMR patients and non-TMR subjects (n = 4) with transfemoral amputations. Shaded regions represent ± 1 SD.

sensors. Inclusion of the EMG signals reduced classification errors rates from 12.9% to 1.8% (Hargrove et al. 2013). Native residual thigh muscles provided improvements of 25%, and the signals measured from over the reinnervated muscle provided an additional improvement of 19%. More important, the patient was able to move spontaneously about the laboratory using the real-time control system with few misclassifications.

10.3 Advanced Prosthetic Systems

10.3.1 Advanced Electromechanical Arm Systems

TMR results in an increase in the number of prosthetic movements that can be intuitively controlled by high-level upper limb amputees. This innovation is enabling better control of new multi-degree-of-freedom prosthetic hands and wrists and is motivating the industry to build devices with more control inputs. There has also been huge recent growth in the research and development of more complex arm systems with an increasing number of functions. One of the first highly articulated arm systems was the Rehabilitation Institute of Chicago's Six-Motor Arm (Miller et al. 2008). It was assembled from a variety of components manufactured worldwide and has been a valuable asset for research on the controllability of multifunctional prosthetic arms. Unfortunately, the system was not robust enough for individuals to use at home.

The Revolutionizing Prosthetics program undertaken by DARPA from 2005 to 2009 resulted in the development of two advanced, modular research arm prototypes. DEKA Research and Development Corporation designed and built a 10-degree-of-freedom modular arm system that could be fitted at the shoulder-disarticulation, transhumeral, or transradial levels. TMR amputees fitted with early DEKA arm prototypes demonstrated excellent control during limited trials. The 22-degree-of-freedom Modular Prosthetic Limb system designed by the Johns Hopkins Applied Physics Laboratory, which included individually actuated fingers, was an even more ambitious undertaking. This device provides unprecedented mechanical agility and was designed to use EMG, peripheral nerve, or cortical signals as inputs. However, due to its complexity and cost, it is unlikely to be made commercially available in its current form.

The size and weight of the prosthesis is a critical issue for prosthesis users. Currently, most prostheses are made for the 50th percentile male; thus they are too large for more than 70% of people. Current prostheses are also relatively heavy; 88% of non-prosthetic users cited weight as a primary reason they choose not to wear a prosthesis (Biddiss and Chau 2007). The Rehabilitation Institute of Chicago has designed a modular six-degree-of-freedom prosthetic limb suitable to fit the 25th percentile female (which also equals the 50th percentile 12-year-old child). The device can be made longer and have larger foam covers to fit larger people. This prosthesis will be suitable for transradial and higher levels of amputation and will support pattern recognition control algorithms. The University of New Brunswick is also seeking to develop a small, low-cost pattern recognition–controlled prosthetic hand, and Vanderbilt University is developing a unique, lightweight prosthetic hand. In contrast to the limbs developed as part of the Revolutionizing Prosthetics program, these systems are designed to support fewer degrees of freedom but will be lighter and more affordable. Otto Bock has released the DynamicArm® TMR (Otto Bock HealthCare GmbH, Duderstadt, Germany), a version of its DynamicArm® prosthesis specifically designed to facilitate TMR fittings. This arm allows for the simultaneous control of the elbow and an Otto Bock terminal device but does not yet support pattern recognition. Several multi-function hands have recently entered the marketplace. These hands support multiple grasps but often the thumb needs be manually repositioned.

Recent technological advancements in actuators, transmissions, and energy storage have resulted in the technical and commercial feasibility of mechanically powered prosthetic legs. Several new motorized leg systems are in various stages of development (Holgate et al. 2008; Martinez-Villalpando et al. 2008; Varol et al. 2010; Lawson et al. 2011). These devices use a control strategy similar to that of a microprocessor-controlled mechanically passive prosthesis: high-level state-based controllers interpret signals recorded from mechanical sensors embedded in the prosthesis or from an orthosis placed on the sound limb. The current state of the device then provides control information to lower-level position, force, torque, or impedance controllers to ensure that the actuator behaves properly. The ability to generate positive mechanical power greatly increases the number of locomotion modes that can be restored to individuals; however, a robust neural interface is required before these devices can be used to their fullest potential.

10.3.2 Advanced Prosthetic Interfaces

Innovative socket designs, as described in Chapter 6, are often required for individuals who have had TMR surgery. It is critical that electrodes maintain excellent contact with the user's skin even if the underlying muscle moves as a result of strong contractions. Comfortable off-the-shelf silicone gel liners have been used for individuals with lower limb amputation for many years but have not been used extensively for people with upper limb amputation because it is challenging to embed electrodes within the liner material. Recently, conductive fabric textile electrodes have been incorporated into the design of silicone gel liners suitable for transhumeral, transradial, and lower limb amputees. These liners maintain excellent skin-electrode contact and result in a comfortable fitting for the user. In some circumstances, off-the-shelf liners may be inadequate, in which case customized silicone liners that are unique to each individual can be used, although these would be more expensive.

Even the most comfortable sockets are still suspended from the body via soft tissue. Consequently, the prosthesis feels like dead weight and feels heavy to the user, even if it is actually lighter than the missing limb. The use of percutaneous skeletal attachments, in which the prosthesis is directly attached to the bone in the residual limb (also known as *osseointegration*), is a promising area of research. The Branemark group (Branemark et al. 2001; Palmquist et al. 2008) have developed a surgical technique and hardware system that has now been implanted in more than 100 people around the world. (The system is not currently approved by the Food and Drug Administration for use in the United States.) There are two major safety concerns associated with percutaneous skeletal attachments: (1) loosening or breakage of the osseointegrated implant, and (2) infection of the skin and surrounding tissues at the skin-implant interface where the implant protrudes through the skin. Both of these issues are areas of active, ongoing research (Sullivan et al. 2003; Pitkin 2008; Williams et al. 2010). Progress in these areas will likely make direct skeletal attachment a more viable option for individuals with major limb amputation.

10.3.2.1 Intramuscular EMG Systems and Associated Algorithms

Surface EMG signals are convenient to work with because they are obtained noninvasively and require relatively inexpensive hardware (Aggarwal et al. 2008); however, there are limitations to their use. Surface EMG signals can only be recorded from superficial muscles, and the layer of soft tissue and skin between the muscle and the recording electrode produces weaker signals that are often a combination of signals from neighboring muscles (i.e., muscle cross talk).

Intramuscular measurement of EMG signals results in high-quality signals with reduced cross talk. Intramuscular EMG signals have traditionally been less convenient to work with due to the invasive nature of the necessary percutaneous wires. Technological advances in electronics and packaging (e.g., Implantable Myoelectric Sensors) (Weir et al. 2009) make it reasonable to assume that implantable recording systems will be available in the near future. These systems will use wireless telemetry to transfer recorded EMG data out of the body, eliminating the percutaneous connection. Intramuscular EMG recordings will expand the options to extract control information, improve the performance of the amplitude-based control algorithms, and make the configuration of algorithms much simpler. Reinnervated control sites are spatially separated during TMR surgery so that independent control sites may be obtained on the skin surface; however, it can still be challenging to locate the ideal electrode placement to obtain signals free of cross talk. Intramuscular electrodes will reduce the effects of muscle cross talk, making it easier to isolate signals from independent muscles and configure activation thresholds for prosthesis control.

The use of intramuscular EMG signals also enables use of a promising method to achieve simultaneous independent control of multiple prosthetic movements for both TMR and transradial patients. This method is based on the hypothesis that the human central nervous system does not directly coordinate the activation of individual muscles, but instead directs the coactivation of sets of muscles—called *muscle synergies* (Lee 1984; MacPherson 1991; Tresch et al. 1999). A muscle may belong to multiple sets, or synergies, and the overall activation state of a muscle is the result of the combined activations of all synergies of which that muscle is a member. Muscle synergies have been successfully identified from EMG recordings using a variety of blind source separation algorithms (Tresch et al. 2006). Blind source separation refers to the recovery of source signals (e.g., synergies) that have been mixed to form an EMG signal, without knowing how the information was mixed. A variety of algorithms with various simplifying assumptions have been explored to identify muscle synergies, including independent component analysis, which assumes that source signals are independent, and non-negative matrix factorization, which assumes that source signals are non-negative (Tresch et al. 2006). Both of these signal processing techniques may prove useful in decoding intramuscular EMG signals.

10.4 Conclusion

TMR has resulted in immediate clinical benefits and has provided diverse research and development opportunities that have benefitted people with amputations whether or not they have had TMR surgery. For example, many improvements in pattern recognition have been developed and first tested on TMR patients, but non-TMR transradial amputees certainly stand to benefit from this research. TMR surgery may readily be extended to other amputee populations, including those with transradial and lower limb amputations. Pattern recognition allows control over many more degrees of freedom than is possible using conventional EMG amplitude methods. Evolving technologies, including implantable EMG sensors and direct skeletal attachment, are ideally suited to complement TMR and will provide amputees with unprecedented control over existing and advanced myoelectric prostheses.

References

Aggarwal, V., S. Acharya, F. Tenore, et al. 2008. Asynchronous decoding of dexterous finger movements using M1 neurons. *IEEE Trans Neural Syst Rehabil Eng* 16 (1):3–14.

Agnew, S. P., A. E. Schultz, G. A. Dumanian, and T. A. Kuiken. 2012. Targeted reinnervation in the transfemoral amputee: a preliminary study of surgical technique. *Plast Reconstr Surg* 129 (1):187–194.

Au, S., and H. Herr. 2008. Powered ankle-foot prosthesis. The importance of series and parallel motor elasticity. *IEEE Robotics and Automation Magazine* 15(3):52–59.

Biddiss, E., and T. Chau. 2007. Upper-limb prosthetics: critical factors in device abandonment. *Am J Phys Med Rehabil* 86 (12):977–987.

Branemark, R., P. I. Branemark, B. Rydevik, and R. R. Myers. 2001. Osseointegration in skeletal reconstruction and rehabilitation: a review. *J Rehab Res Dev* 38 (2):175–181.

Dillingham, T. R., L. E. Pezzin, and E. J. MacKenzie. 2002. Limb amputation and limb deficiency: epidemiology and recent trends in the United States. *South Med J* 95 (8):875–883.

Hargrove, L., K. Englehart, and B. Hudgins. 2007. A comparison of surface and intramuscular myoelectric signal classification. *IEEE Trans Biomed Eng* 54:847–853.

Hargrove, L., A. Simon, R. Lipschutz, S. Finucane, and T. A. Kuiken. 2011. Real-time myoelectric control of knee and ankle motions for transfemoral amputees. *JAMA* 305 (15):1442–1444.

Hargrove, L., A. Simon, R. Lipschutz, S. Finucane, D. Smith, and T. A. Kuiken. 2013. Robotic leg control with EMG decoding by an amputee with nerve transfers. *New Engl J Med*. Accepted.

Holgate, M. A., J. K. Hitt, R. D. Bellman, T. G. Sugar, and K. W. Hollander. 2008. *The SPARKy (Spring Ankle with Regenerative Kinetics) project: choosing a DC motor based actuation method*. Paper read at the second IEEE RAS & EMBS International Conference on Biomedical Robotics and Biomechatronics, 19–22 Oct. 2008.

Huang, H., T. Kuiken, and R. Lipschutz. 2009. A strategy for identifying locomotion modes using surface electromyography. *IEEE Trans Biomed Eng* 56 (1):65–73.

Huang, H., F. Zhang, L. J. Hargrove, D. Zhi, D. R. Rogers, and K. B. Englehart. 2011. Continuous locomotion-mode identification for prosthetic legs based on neuromuscular mechanical fusion. *IEEE Trans Biomed Eng* 58 (10):2867–2875.

Huang, H., P. Zhou, G. Li, and T. A. Kuiken. 2008. An analysis of EMG electrode configuration for targeted muscle reinnervation based neural machine interface. *IEEE Trans Neural Syst Rehabil Eng* 16 (1):37–45.

Kuiken, T. A., G. A. Dumanian, R. D. Lipschutz, L. A. Miller, and K. A. Stubblefield. 2004. The use of targeted muscle reinnervation for improved myoelectric prosthesis control in a bilateral shoulder disarticulation amputee. *Prosthet Orthot Int* 28 (3):245–253.

Kuiken, T. A., G. Li, B. A. Lock, et al. 2009. Targeted muscle reinnervation for real-time myoelectric control of multifunction artificial arms. *JAMA* 301 (6):619–628.

Lawson, B. E., H. A. Varol, and M. Goldfarb. 2011. Standing stability enhancement with an intelligent powered transfemoral prosthesis. *IEEE Trans Biomed Eng* 58 (9):2617–2624.

Lee, W. A. 1984. Neuromotor synergies as a basis for coordinated intentional action. *J Motor Behav* 16 (2):135–170.

Li, G., A. E. Schultz, and T. A. Kuiken. 2009. Quantifying pattern recognition-based myoelectric control of multifunctional transradial prostheses. *IEEE Trans Neural Syst Rehabil Eng* 18 (2):185–192.

MacPherson, J. M. 1991. How flexible are muscle synergies? In *Motor Control: Concepts and Issues*, edited by D. R. Humphrey and H.-J. Freund. West Sussex, England: John Wiley & Sons Ltd.

Martinez-Villalpando, E. C., J. Weber, G. Elliott, and H. Herr. 2008. *Design of an agonist-antagonist active knee prosthesis*. Paper read at the second IEEE RAS & EMBS International Conference on Biomedical Robotics and Biomechatronics, 19–22 Oct. 2008.

Miller, L. A., R. D. Lipschutz, K. A. Stubblefield, et al. 2008. Control of a six degree of freedom prosthetic arm after targeted muscle reinnervation surgery. *Arch Phys Med Rehabil* 89 (11):2057–2065.

ÖSSUR. *The POWER KNEE™*. The POWER KNEE ed: http://bionics.ossur.com/Products/POWER-KNEE/SENSE.

Palmquist, A., T. Jarmar, L. Emanuelsson, R. Branemark, H. Engqvist, and P. Thomsen. 2008. Forearm bone-anchored amputation prosthesis: a case study on the osseointegration. *Acta Orthop* 79 (1):78–85.

Pitkin, M. 2008. One lesson from arthroplasty to osseointegration in search for better fixation of in-bone implanted prosthesis. *J Rehab Res Dev* 45 (4):vii–xiv.

Scheme, E., and K. Englehart. 2011. EMG pattern recognition for the control of powered upper limb prostheses: state-of-the-art and challenges for clinical use. *J Rehab Res Dev* 48 (6):643–659.

Simon, A., L. Hargrove, B. A. Lock, and T. A. Kuiken. 2011a. A decision-based velocity ramp for minimizing the effect of misclassifications during real-time pattern recognition control. *IEEE Trans Biomed Eng* 58 (8):2360–2368.

Simon, A., L. Hargrove, B. A. Lock, and T. A. Kuiken. 2011b. Target Achievement Control Test: evaluating real-time myoelectric pattern-recognition control of multifunctional upper-limb prostheses. *J Rehab Res Dev* 18 (6):619–628.

Simon, A., B. A. Lock, and K. Stubblefield. 2012. Patient training for functional use of pattern recognition-controlled prostheses. *J Prosthet Orthot* 24 (2):56–64.

Sullivan, J., M. Uden, K. P. Robinson, and S. Sooriakumaran. 2003. Rehabiliation of the trans-femoral amputee with an osseointegrated prosthesis: the United Kingdom experience. *Prosthet Orthot Int* 27 (2):114–120.

Sup, F., A. Bohara, and M. Goldfarb. 2008. Design and control of a powered trans-femoral prosthesis. *Int J Robot Res* 27 (2):263–273.

Tkach, D., A. Young, L. H. Smith, and L. Hargrove. 2012. Performance of pattern recognition myoelectric control using a generic electrode grid with targeted muscle reinnervation patients. *Conf IEEE Eng Med Biol Soc* Aug 2012:4319–4323.

Tresch, M. C., V. C. K. Cheung, and A. d'Avella. 2006. Matrix factorization algorithms for the identification of muscle synergies: evaluation on simulated and experimental data sets. *J Neurophysiol* 95 (4):2199–2212.

Tresch, M. C., P. Saltiel, and E. Bizzi. 1999. The construction of movement by the spinal cord. *Nat Neurosci* 2 (2):162–167.

Varol, H. A., F. Sup, and M. Goldfarb. 2010. Multiclass real-time intent recognition of a powered lower limb prosthesis. *IEEE Trans Biomed Eng* 57 (3):542–551.

Weir, R. E. F., P. R. Troyk, G. A. DeMichele, D. A. Kerns, J. F. Schorsch, and H. Maas. 2009. Implantable myoelectric sensors (IMESs) for intramuscular electromyogram recording. *IEEE Trans Biomed Eng* 56 (1):159–171.

Williams, D. L., R. D. Bloebaum, J. P. Beck, and C. A. Petti. 2010. Characterization of bacterial isolates collected from a sheep model of osseointegration. *Curr Microbiol* 61 (6):574–583.

Williams, T. W. 2004. Control of powered upper extremity prostheses. In *Functional Restoration of Adults and Children with Upper Extremity Amputation*, edited by R. H. Meier and D. J. Atkins. New York, NY: Demos Medical Publishing.

Young, A., L. Hargrove, and T. A. Kuiken. 2012a. Improving myoelectric pattern recognition robustness to electrode shift by changing interelectrode distance and electrode configuration. *IEEE Trans Biomed Eng* 59 (3):645–652.

Young, A., L. H. Smith, E. Rouse, and L. Hargrove. 2012b. *A new hierarchical approach for simultaneous control of multi-joint powered prostheses.* Paper read at the fourth IEEE RAS & EMBS International Conference on Biomedical Robotics and Biomechatronics, June 24–27.

Zhou, P., M. M. Lowery, K. B. Englehart, et al. 2007. Decoding a new neural-machine interface for control of artificial limbs. *J Neurophysiol* 98 (5):2974–2982.

Ziegler-Graham, K., E. J. MacKenzie, P. L. Ephraim, T. G. Travison, and R. Brookmeyer. 2008. Estimating the prevalence of limb loss in the United States: 2005 to 2050. *Arch Phys Med Rehabil* 89 (3):422–429.

Index

Printed and bound by CPI Group (UK) Ltd, Croydon, CR0 4YY

21/10/2024

01777112-0003